THE HUMEM STATE

The Emergence and
Establishment of Our
Extended Presence

Alan Brook

Humemity
ISBN-13: 978-0692251621
ISBN-10: 0692251626

ABOUT THE AUTHOR

Alan Brook has spent most of his career developing digital communications systems in both startups and major tech companies. He has dedicated the last few years to developing a fundamentally new approach to the future of individuals' digital extended presences. *The Humem State* is part of this ongoing endeavor. He holds degrees in physics and mathematics. He is an avid runner and kayaker, and lives in Silicon Valley, California.

CONTENTS

PART III ESTABLISHMENT

PART IV OUTCOMES AND OPPORTUNITIES

PROLOGUE

Our personal spheres of influence extend far beyond our bodies. This *extended presence (EP)* includes the things we create and our effects on objects and other people—that is, our imprint on the world exterior to ourselves.

In the past, only celebrities—such as state and military leaders, writers, and actors—possessed widespread and enduring EPs, which were mostly limited to their public personas. But in recent times, digital information technology and related developments, such as social media, have resulted in the emergence and rapid growth of an extensive and very personal form of EP for most modern people. This new EP is becoming an extension of our selves and a primary means of communication in both the private and professional realms of our lives. It is an increasingly inseparable part of who we are and what we do.

Despite the personal EP's astounding progress, it suffers from multiple deficiencies: it is often fragmented and transient, and is subject to infringements of privacy and control. Yet, most of our EP's sources, and our underlying motives for its creation, are intricately interconnected, enduring, and very private.

As personal data applications become much more powerful, useful, and omnipresent, our dependence on them will grow until we become virtually helpless without them. (As could be said of electric-powered technology today.) When our EPs contain a record of virtually all our actions and interactions, nearly every word we've ever uttered or heard, an account of almost all our memories and knowledge, and a log of every beat of our hearts or fluctuation in our blood-sugar levels, these technologies' potential for good will become immense. But, as with most powerful devices, the potential for their misuse will also rise. On current trajectories, as information technologies advance, individuals continue to lose ever more privacy and control of their EPs.

When we consider pressing problems in technology or medicine, we often assume that our "best minds" are working on finding solutions. However, with regard to many of the deficiencies of the personal EP, quite the opposite is true. The entities with the greatest power and influence in this domain, including corporations and nation-state institutions, often have interests that are not well aligned with, or are even starkly opposed to, those of individual data creators and consumers. And it is these same entities that employ and direct the efforts of the vast majority of our best technologists.

Consequently, with the exception of a few individuals and academics, and the staff of some small organizations, the best minds in technology are not engaged in solving these deficiencies, but are instead intently focused on creating a next generation of applications that will actually exacerbate these EP problems. Almost invariably, this is not due to any malevolence on their part but is simply a result of where the most interesting and financially rewarding employment is to be found.

Current attempts at correcting these EP shortcomings, such as the meager advances in data-privacy legislation, do not address their underlying causes and are little more than Band-Aids on a rapidly worsening affliction.

As our modern EP becomes ever more essential to our lives and an integral part of our personas, many of its needs begin to parallel our own. Thus, it requires rights and protections, including the right to exist in security for the duration of its natural lifespan (which may be much longer than ours) and the right to privacy. On further examination, it soon becomes evident that these basic rights are dependent on a broad set of additional ones, such as the right to freedom from being owned by others and the right to economic autonomy.

Advanced nation-states, with their laws and conventions that regulate relationships among individuals, societies, and states, have usually provided the optimal frameworks for ensuring the well-being of individuals. They establish constitutions, recognize citizens' rights, and pass laws. They institute law enforcement services, safety standards, and health and education systems, and build numerous other physical and organizational infrastructures to serve their citizens' needs. Our EPs require similar apparatuses to ensure their integrity and well-being. The resulting EP empowerment and protections are of crucial consequence for many of our "real-world" concerns, such as our personal information technology and financial systems' security, our credit ratings, our personal safety and privacy, and our professional and social reputations.

In this book, I argue that existing systems in nation-states, or incremental adaptations of existing systems, are fundamentally incompatible with both the short- and long-term needs of our EPs. Traditional protectors, such as law enforcement agencies, are increasingly ineffectual in the face of powerful commercial forces or international cybercriminals. Regional and even national security mechanisms are confined by the limits of their jurisdictions and thus are inherently ill suited to confront the nebulous, anonymous, and often-global threats to the EP's integrity. Even the relevant legislation (where it exists) is inadequate and lags far behind technological progress. But perhaps most significantly, our own governmental agencies—the supposed guardians of our freedoms—are themselves frequently responsible for major transgressions on our privacies and other fundamental EP rights.

Consequently, a comprehensive solution to these problems cannot come in the form of minor adjustments to current systems; it cannot be established on the decaying foundations of outdated institutions or on foundations that were designed for completely different purposes. A fundamentally new approach is required.

I seek to convince you, the reader, that to allow our EPs, as extensions of our beings, to become all that they can be and everything that we wish them to be, it is essential to consolidate their parts into conceptually unified bodies, each representing an individual. Moreover, these resulting entities need to be formally—legally and economically—independent from the people from whom they are derived. To achieve this, they should be established and recognized in dedicated state-like structures in which they possess a suitable set of citizen-like rights. To reflect the parallelism between these formalized consolidations of our EPs in their hosting environments and people in nation-states, I call the new entities *humems* in *humem-states*.

Only a dedicated state-like system can ensure the continuance of environment and consistency of purpose that are needed for our humems' long-term welfare; only a robust and well-funded state-like administration can counterbalance the EP related interests of corporations and nation-states, and counteract the EP detriments of nefarious individuals and groups; and only a customized state-like economic and legal environment can allow us and our humems to own and benefit from the immense value that they generate.

This book is an outgrowth of a number of years of practical preparation for building the foundation of a dedicated state-like system within existing jurisdictions. Underlying the vision is a real and concrete infrastructure designed by mainstream technologists and legal, financial, and taxation experts. Yet, state building—even in this ultra-modern form—requires the common will and coordinated effort of a broad assemblage of individuals and institutions. Here is the primary purpose of this work: to create a coherent worldview and terminology that will facilitate the communal actions needed to make the humem system a reality.

Overview

The corpus of this book is organized into four parts:

Part one, "Perspective," builds the foundation for the new ideas to be presented in the subsequent chapters. By examining a number of familiar, interrelated concepts in a fresh way, this section establishes a terminology to describe our extended presence. In addition, it portrays the EP as a much more fundamental and vivacious entity than most people presently perceive it to be—an entity that even appears to possess a degree of self-volition.

Part two, "Emergence," discusses the seemingly spontaneous emergence of discrete bodies formed from consolidations of the personal EP, the initial instances of which I call *proto-humems* (early humems). It describes their characteristics, capabilities, and wide-ranging potentials, and shows how rudimentary manifestations of these bodies are already observable in many places around us. This section examines several of the essential advantages gained through the aggregation of one's EP into a conceptual unit—one's personal humem—and then introduces some of the predominant interactions and relationships between humems and people.

Part three, "Establishment," contains the core prescriptions, or calls-to-action, of the book. It argues that a natural outcome of humems' facilitated emergence is the imperative to provide them with rights comparable to those of state citizens secured by government-like administrations. It outlines the various downsides of current approaches, and proposes the type of administration and economic infrastructure necessary for the welfare and regulation of the humem entities.

Part four, "Outcomes and Opportunities," explores a number of the remarkable consequences and potential benefits of established humems. It goes on to speculate how humems may fulfill some of our most fundamental needs and desires in the future. Finally, it addresses some of the predominant concerns people may have with regard to these rapidly proliferating entities.

The epilogue summarizes the book's main precepts and vision and, in conclusion to arguments set forth in earlier chapters, indicates how the notion of a union composed of a person and humem provides a more consistent way of perceiving the fundamental unit of future society.

Scope, Audience, and Expectations

This work is intended to present a new worldview to a broad audience, and to impart a sense of structure and a feeling for the interactions between the parts of a comprehensive system. Such a synthesis sometimes comes at the expense of details and stringent technical arguments. Although I treat the core issues in considerable depth, in some of the more peripheral areas I have condensed the coverage of complex topics to a few lines of broad assertions or observations, which are intended mainly as a basis for future work and also to complete the outline of the big picture.

Among others, I have envisioned audiences as diverse as information technologists, legislators, and those general readers who are curious and concerned about the future of our personas under the influence of our new technologies. In some places, this has resulted in a somewhat more drawn out coverage than would have been necessary had the book been dedicated to a specific category of reader. However, in most cases, I have tried to remain concise, with the expectation that the modern, Internet-connected reader is easily able to supplement the material where needed.

For example, where I assert that digital data suffers from a number of acute durability or longevity problems, most information technology professionals would readily understand what I mean. If asked to do so, they would easily be able to list some of the most common deficiencies. But a legislator or lawyer may be much less adept at this exercise. Yet, had I attempted to provide a review of this and similar subjects, I would have risked both boring the technologist and doubling the length of the book. Thus, I have presumed that the reader is able to enter a term such as "digital data longevity" into an Internet search engine, where they will be presented with a variety of additional reading options.

Likewise, when advocating for our personal EPs' rights by comparing their current vulnerabilities to those of disadvantaged people in the past, I have not provided an overview of the circumstances prevailing during those unfortunate historical episodes. Instead, I have assumed that the technologists who might have missed or forgotten their history lessons can similarly easily access such information.

I expect that, at least initially, some readers will get the impression that I am describing some sort of imaginary system, or what is sometimes called a virtual world; this would be a fundamental misunderstanding. I am dedicated to describing an entirely practical system that is no less tangible than those things most of us deem real in everyday life. We need no new technologies or utopian developments in our cultural or legal systems to recognize these emerging entities and to establish systems for fostering their development. Throughout this book, I have endeavored to interleave the presentation of the vision with the practical details of its establishment. This approach is intended to give you a continuous and intuitive sense of the reality and feasibility of the proposed system.

The entities that I describe in this book are becoming an integral part of our lives. They will be born with us, accompany us throughout life, and persist beyond. Many cultures teach very young children idealistic worldviews like religion, politics, and the meaning of true and good and bad. It is my hope and expectation that the core ideas of this work, if presented concisely and in the appropriate context, would also be understandable to young readers who, with their inherent plasticity of thought, may perhaps comprehend them best.

Value does not imply perfection. In this work, I first propose some novel ways of considering familiar things. Then, I introduce some fundamentally new viewpoints and advocate the creation of dedicated large-scale institutions. In today's world of specialists, no one person is a historian, technologist, sociologist, theologian, economist, politician, anthropologist, and legislator. I do, nonetheless, touch on a broad swath of disciplines, including those of the professions listed above. Thus, without doubt, I will get some of the particulars wrong. This is especially a risk where, by extrapolating the essentials of the vision, I venture to forecast some of the implications and outcomes of these developments.

I fully believe, however, that a considerable measure of audacity is required to clearly present the worldview advocated in this book. My primary intent is to seed thought as a prelude to understanding these potentials and creating a common vision—an essential first step toward rallying the communal actions required to facilitate the emergence and establishment of our personal extended presence. As long as the book's core ideas are viable and valuable, even if some details do not stand up to future tests, this quest will have still been well worth the effort.

Terminology and Dissipating Metaphors

Sometimes, when venturing into unfamiliar domains, we have to first imagine answers before we can articulate our questions. Then, by employing the instruments of the new frameworks and terminologies produced by these imaginings, we can begin to more methodically validate our premises.

As we do this, and continue to stretch our imaginations, we may find that we also need to stretch our vocabulary. Frequently, a careful and qualified reapplication of existing terms is appropriate; but occasionally, really new things deserve really new names. As evidenced by the title of this book, I am not completely averse to the latter option. However, in most cases I will be reusing existing expressions, because, once their contexts are properly clarified, these terms are still entirely valid for our current purposes.

In the study of a new system, the naming itself can be crucial to understanding mechanisms and interactions among the parts. In particular, it is often valuable to create a broad and consistent metaphor that encompasses all the system's known properties. Such a metaphor, if it can be found, can help establish both a framework for more focused analysis and a terminology for more coherent discussion. In certain circumstances, such metaphors may result in more potent outcomes: seemingly unrelated systems that possess analogous external characteristics may subsequently be seen to share common underlying mechanisms; and then, a thorough understanding of one system may reveal the workings of the second, less-known system.

Occasionally, we may even discover that what we thought was only comparable in name and behavior is really one and the same. In such a case, the metaphor dissipates and transforms into a true and exact description.

I introduce the core concepts of this book by utilizing a number of familiar metaphors. To date, in discussions with others, I have found this to be by far the most effective method of conveying these ideas. I anticipate that, by the end of this book, you will easily be able to judge for yourself the aptness of the naming, and to what degree certain core metaphors have dissipated and a proper and coherent terminology has emerged.

Exploring Together

Most of our advances in understanding are attained in modest increments, building on well-established reference points—the things we know or believe we know well. Frequently, when given an incomplete or brief glimpse of a new object, it seems that our minds cannot tolerate the vagueness; they are compelled to fill the void. They spontaneously create a model, or a best approximation, of how the object might appear if we were able to examine it more completely; where patterns are not apparent from without, we create them from within.

Such contrived images invariably resemble the aforementioned reference points. Thus, if it happens that the object we glimpsed is actually similar to something we know, we may be somewhat successful in our predictions. If, however, it is very different from anything we are familiar with, then our imaginings may deviate drastically from reality. In such circumstances, the cursory glimpse places us further from the true picture than if we had no knowledge of the object whatsoever.

Such is the nature of my challenge as I introduce this book. At the time of this writing, the concepts contained here are starkly different from most readers' existing reference points. Consequently, a brief preview, such as this introduction, may induce you to invoke a derivative of a familiar image that is very different from what I intend to present. Thus, I ask you to take this overview for what it is—a very partial sketch of something much larger—and, as far as you are able, to maintain an open mind on what is to follow.

When I deliberated the mode of presentation, I considered a number of standard approaches. One option was to set forth the structure and premises of the new system, to show what it can do and will be, and then attempt to demonstrate the validity of these claims—akin to a mathematician first stating a theorem and then offering the proof, or an attorney in a court of law, alleging a perception of truth as an account of a sequence of events and only later presenting the evidence and bringing in the witnesses.

Instead, I have taken a different approach, by reciting the story in a way that resembles the sequence of its natural emergence. With the exception of what I have revealed in the prologue, I have attempted to create a mission of co-discovery with you, the reader. I have done this by recreating my own journey, albeit in an abbreviated and more structured manner. In this way, the picture emerges only gradually, and it takes a bit longer to comprehend the overall structure. Such is, however, the essence of exploration. Your sense of confidence and familiarity from having participated in the gradual exposure of this worldview is the best outcome that I envision. My ardent hope is that, having journeyed once with me, you will be able to independently retrace your steps and then confidently venture alone into adjacent areas that we did not traverse together.

I have endeavored to present the process and outcomes as transparently as I can. Nowhere are you required to blindly trust my assessments. Neither are you asked to accept the calculations, interpretations, conclusions, or predictions of others—however expert they may be. Even as we explore together, ultimately, you stand alone in your judgment of what you observe.

Embarking on this journey of discovery, you, as an explorer, are not required to take any wild leaps into the unknown. But, for a short time only, I challenge you to loosen your grip on your preconceptions. Then, as you gaze forward through the dissipating dust of old impossibilities, you may catch your first glimpse of something new and wonderful.

Part I
PERSPECTIVE

CHAPTER 1

OUR EXTENDED PRESENCE

Beyond Our Bodies

For the purpose of our inquiry let's start from first principles and conceptually divide a person's presence in the world into two parts. First, their bodily presence—that which is contained within the physical confines of their body; and second, their *extended presence* (EP)—their imprint or influence on the world exterior to their body.[1]

We can think of a person's EP as the compendium of their influences on their surroundings: all the changes caused to the external world as a result of their existence. In this context, "surroundings" can be understood in the most general sense possible—that is, everything external to the person's body at any given point in time.

An important type of EP relates to our influences on other people, in particular, the changes made to other people's brains as a result of our being. We can call this kind of EP a *bio-EP*, because it is carried by a biological entity. When we interact with other people, via their senses,

---

[1] Notably, the designation of the extended presence versus the bodily presence is relative to a specific individual. In particular, parts of one person's extended presence may reside in another's bodily presence. As we progress in our discussion, this will become clearer.

we cause potentially long-lasting, or even permanent, changes to their brains; we leave our mark on their brains. This is a concrete phenomenon that manifests in many ways—most simply in those people's ability to recognize us at a later time.[2] The fact that they can identify, from a distance, our face, voice, or gait, means that our previous interactions resulted in a unique imprint on their brains. They now *know* us. In addition, their memory allows them to replicate some of our attributes. For example, with varying levels of fidelity, they can repeat things we have said, perhaps sketch a likeness of us on paper, mimic a characteristic movement of ours, or hum a tune that we composed. Consequently, those who know us can also propagate our bio-EP into other people's minds.

The magnitude of this type of EP is reflected in metrics such as the number of people who know us, and how well they know or have been impressed by us—that is, the scope of our social or professional imprint. In general, we can think of our EP existing in two dimensions: its "width," or its extent at any given time; and its "length," or the duration for which it persists.

Most social concepts can be described in terms of EP. For example, a famous person is someone who has an extensive EP and is considered popular to the extent that others wish to increase the scope of that person's EP in their minds. Someone who is intensely disliked or feared can also have a very large EP; we typically call that person "notorious." In such a case, others may strive to increase that individual's EP in a "know thy enemy" sense. We can consider "reputation" to be a measure of the quality of one's EP, as in a good or a bad reputation.

In addition to the imprint that one leaves on another's mind, there are other types of bio-EP. An obvious example is one's biological offspring. Also, the results of a surgical procedure, haircut, or tattoo are bio-EPs of the surgeon, hairstylist, or tattoo artist, respectively. Each has influenced, or left a mark on, another person's body. Similarly, one's

---

[2] Clearly, our pets also can recognize us. Thus a similar bio-EP resides in animals as well.

effects on other living things—animals and plants—are also components of one's bio-EP.

Another category of EP is composed of our influences on the inanimate environment. We can call this our *abiotic-EP*.[3] Examples of this include the house I build, the autograph I sign, the software I code, the fire I light, and the book I write. Also included here are pictures of me, writings about me, institutional records relating to me, recordings of my voice, and so on. For simplicity, we can consider all EP that is not bio-EP to be abiotic-EP.

For most of us who live in technological civilizations, digital data has recently become a dominant component of our abiotic-EP. This pertains to our vast digital footprint, which includes both data that we create intentionally and an array of other forms of data that are generated indirectly as a result of our existence.

Think about yesterday, for example. Dissect it into small time segments and consider if and how, in each of these periods, you were actively creating data, or less directly, how data was being created and influenced by your presence or actions. When using electronic devices, most of us are aware, to a greater or lesser extent, of how we are generating data. However, we are typically less mindful of the many more nuanced ways in which we leave an imprint on the digital world. For example, by "passively" carrying a mobile communications device, we continuously produce information relating to our location. (This may or may not be recorded, depending on the configuration of the device, the policies of the communications provider, the laws of the jurisdiction in which we reside, and numerous other factors.) Additionally, there are even more imperceptible processes by which we are creating data. By simply walking down the street we leave a digital footprint in surveillance cameras. Or, when we turn on the water faucet in our kitchens or switch on the lights—actions we don't normally think of as creating

---

[3] *Abiotic* means non-biotic or non-biological. Other options for the naming of this kind of EP could have been *inert-EP* or *inanimate-EP*. I expect that my reluctance to adopt such terms will become clearer once we begin observing the dynamic and lifelike behavior of certain kinds of abiotic-EP.

digital data—we induce changes in the reading of water or electricity meters, which in turn, feed into a number of computing and data storage systems. These include the utility provider's systems when the meters are read and the bill is created, our communications service provider's systems and our personal computing devices when the utility bill is communicated by email, our banks' or credit card providers' systems when the payment is made, and many others. In fact, when we begin to carefully consider the workings of our modern lives, we find that we are generating or affecting data in almost all our daily actions. Some of these data are transient and inconsequential, but many can persist for indefinite periods and have significant future effects.

Digital data's accelerating growth is the catalyst for the present transformation of our personal EP. Consequently, it is the underlying enabler of much of what we'll discuss in this book. However, just as digital technology was hardly conceivable a century ago, it is quite likely that fundamentally different technologies will emerge and become dominant in the future. Thus, our views of the long-term prospects for our EP should not be overly dominated by the specific nature of current technologies.

A Brief History of Our EP

EPs have changed and expanded over time. For most of history, the EP of the vast majority of people was limited both in extent and duration. In typical circumstances, the memory of a person in the minds of others (the person's bio-EP) was limited mainly to a small number of family members and acquaintances within the same tribe or village. A generation or two following the person's death, virtually none of their bio-EP endured; the bulk was lost when those who directly knew them died as well.

A person's abiotic-EP was composed mainly of the objects they had created, for example, shelters and utensils. Most people had little enduring abiotic-EP. Exceptions existed, such as when those people produced stone arrowheads and cave paintings. But these items were, later on, generally impossible to associate with a specific person. Until fairly recently, the only enduring remnant attributable to an ordinary person was a tombstone with a name, two dates, and possibly a short epitaph. This EP might have been replicated in a church's birth and death register or a municipal archive. In most ways, the EPs of the vast majority of people in the past were not very different in extent and longevity from those of social animals such as elephants or gregarious apes.

In later historical times, the tools of fame and the luxury of legacy emerged. An elite group of people was able to achieve a much greater EP than had previously been possible. Cultural changes related to the establishment of cities and states with their associated hierarchies resulted in environs that facilitated a much larger influence of a few over many. The rich and powerful could employ portrait painters, sculptors, monument builders, and biographers, not only to expand their fame in life, but also to preserve a partial representation of their being after death.

Over time, celebrities such as nobles, politicians, military leaders, philosophers, and writers were portrayed in literature and in fine art. However, their depictions were composed almost exclusively of the best-known facets of their public persona. For example, we may understand a great deal about the theories of certain Classical Greek philosophers yet know little about their personal lives. We may learn the details of a famous general's military exploits but know nothing about his private thoughts. Authors of fiction often left a much deeper legacy of their imaginary characters than of their own lives.

With the spread of literacy, the declining costs of writing material, and the advent of print, there was a moderate increase in the number of people leaving behind longer-lasting imprints. This was still limited, though, to a small part of the total populace. Then, in the 19th century, photography allowed people to generate a more enduring, though still partial, depiction of their physical likeness. Today, many people in the developed world are able to gain a sense of the visual appearance of their great-grandparents by means of a few old photographs. Still, they know little about how their ancestors sounded and appeared when they laughed and smiled.

Over the last century, in industrially advanced societies, technological changes transformed most people's EP. Abiotic media—the media outside of living bodies—started carrying an increasingly larger part of their EP. High rates of literacy and institutional record-keeping resulted in people's lives being documented for a variety of purposes. Governmental registers, licenses, health reports, and other legal and financial tracking left almost no one untouched. Schools exhorted children to document their summer vacations. Some kept diaries. Most people started communicating in writing for personal and professional reasons. In the latter half of the century, many families started owning cameras and began using them with ever-increasing frequency. A lot of this material has survived.

In recent decades, the growth of personal EP has accelerated even more as the world has become digital. Today, in technological societies, the majority of us have an expansive, rapidly growing, and largely involuntary digital footprint. It is increasingly impossible to prevent the proliferation of much of this abiotic-EP without banishing oneself from civilization. With some types of data, particularly the social kind, one can slow or somewhat regulate the creation of one's EP. Most other categories, though, are largely beyond the control of individuals.

Even so, it is clear that the majority of people, especially those born into this digital era, are not overly concerned by the rise of this massive EP. On the contrary, most are cooperating in its creation and are actively competing to cultivate the biggest possible digital EP. They seek to increase the number of their social network friends and followers, the views of their videos, and the visitors to their personal webpages and blogs. Clearly, there is nothing new here: humans, as intensely social creatures, have always striven to increase their visibility and influence in society.

Furthermore, for a reasonable cost and effort, and given the appropriate social context, the majority of us would wish to extend this imprint even beyond the temporal limits of our physical lives. Continuance and legacy are universal human motivations that are consistent over millennia and across cultures. Today most people living in modern societies are leaving, by default, a much more expansive record of their lives than was ever possible in the past. As it becomes increasingly affordable and easy, many are actively attempting to create a permanent EP.

Bodies Optional—Changes in EP Communications

In early times, the EP was communicated mainly from body to body as a one-to-one or one-to-few interaction. A person spoke and gestured while a few others listened and watched. The information was conveyed either by the originator of the EP or via intermediaries. That is, you either listened to your grandmother speaking, or your mom told you what your grandmother had said.

When larger societies formed, and devices such as the stage and stadium enhanced one-to-many communications, the transfer of EP was still restricted by bodily constraints. On one side were the physical stature of the actor and the volume of their voice; on the other side were the size of the audience, the reach of their eyes, and the sensitivity of their ears. These factors limited the EP that could be conveyed by vision or by voice. This was still body-to-body communication; you went to the story because the story could not come to you. In addition, the interaction was time-dependent. You were at the play, or you weren't; there was no replay. The next time—if there were one—would always be different.

Books were the game changer introducing a bodiless and timeless one-to-many method of EP dissemination. Now, the story could come to you whenever and wherever you wished without your having to offer the storyteller dinner. Millions could hear one voice asynchronously, even long after the narrator had departed.

Entertainers, politicians, and athletes still appear on stages and in stadiums. However, microphones and loudspeakers now overcome the constraints of the human voice, and live projections on massive screens mitigate the limitations of body size and eyes, resulting in an enhanced, co-located, time-specific, one-to-many EP transfer. In addition, the event can be broadcast live anywhere thereby overcoming the location-dependence. And the occasion can be recorded thereby removing the time-dependence and enabling unlimited replay long after the originating bodies have moved on or passed away.

Corresponding changes have also occurred in personal EP transfer. A century ago, one could send a letter with a photograph to a distant person, creating a slow, one-to-one, or one-to-few, EP transfer using abiotic-media. This was already a major divergence from the face-to-face, location- and time-dependent, personal bio-EP transfers. However, this kind of communication was not as rich as the traditional face-to-face EP transmission. It was unidirectional (per transfer) and lacked key components of personal interaction, such as body language and the nuances of voice. It could, nevertheless, often leave an enduring record.

The advent of the telephone added real-time bidirectionality and voice to one-on-one distanced interactions but still omitted the visual EP transfer factors. In contrast to the letter and photograph, telephone conversations were not typically recorded and thus did not usually result in the creation of abiotic-EP. The telephone functioned mainly as a transient conduit for bio-EP. Also, significant long-distance telephone costs remained a limiting factor.

Recently, in technological societies, the opportunities for EP transfer have expanded dramatically. Communication methods that overcome most of the old limitations of body, location, and time are readily accessible to all. A kid can video-chat with almost anyone in the modernized world and easily broadcast information to millions of people, all at zero incremental cost. Moreover, it is becoming increasingly feasible to retain a lasting record of all these communications.

A Global Voice for All—EP Democratization

Figure 1 portrays the extended presence of people historically (say, a few centuries ago) and in modern times. It compares the EPs of public figures in each era (the minority) to those of common people (the majority). In this discussion I refer to them as "celebrities" (or "celebs") and "commoners," respectively. These classifications, however imprecise and fluid, are nonetheless useful for our discussion. In the diagrams, the area depicting each of the categories of EP denotes its expanse. These EP existence diagrams are not to scale, yet they do illustrate relative quantities and trends as a function of period.

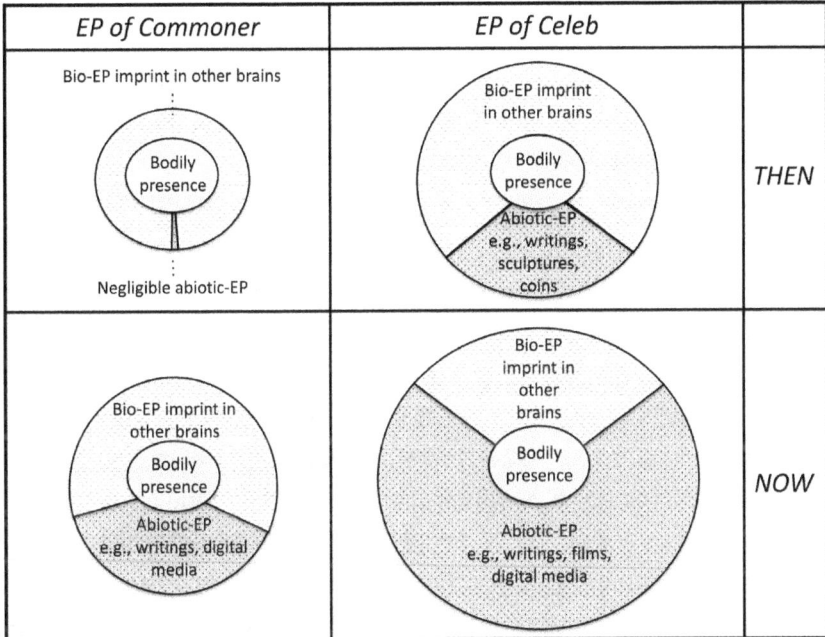

Figure 1: Lifetime EP of celebrity vs. commoner—then and now (comparative—not to scale).

As discussed earlier, most people in the past had virtually no lasting, personal abiotic-EP.[4] Celebs had much greater comparative abiotic-EPs, though they were still modest by current standards. The distinction between the haves and the have-nots was well defined. Today the situation is very different: Everyone in technological societies has a significant and rapidly increasing abiotic-EP. Celebs still have much more, but presently the difference is largely a matter of degree. The ever-growing EP pie is undergoing a dramatic reallocation. Celebs are

---

[4] Clearly, a pyramid laborer may have left a long-lasting imprint on the world. But since it is impossible to attribute the product of their work to the specific person, we can say that this is not *personal* abiotic-EP.

still photographed, filmed, and written about considerably more than others. However, many commoners are also being intensely recorded. If they wish, they can have an almost full-life recording at a fairly modest cost.

Ordinary people now have access to a global voice. In fact, they can rapidly achieve celebrity status in the classical sense: acquiring many followers from the general population spread broadly geographically. This can happen, for instance, when a video posted on a video-sharing site, depicting a common event in an especially engaging way, goes "viral." Likewise, a previously anonymous person who performs some extraordinary or nefarious act can be catapulted to the front page and appear side-by-side with presidents and movie stars.

There is another, more delimited path to a global presence. It relates to one's special or unusual interests—those not typically shared by adjacent people such as family and friends. Through these interests, one can access an audience that is global in a geographical respect, but is typically restricted to a small proportion of the population in any particular place—a long tail of people with common enthusiasms. For instance, an avid interest in falconry or colonial African postage stamps may leave one fairly isolated in a local sense. Worldwide, though, there are likely many others with similar passions.

The figure indicates that modern celebrities' abiotic-EPs, which consist mainly of digital media, may already exceed their bio-EPs. To get a feel for this disparity, consider that we hold much more of celebs' EPs in abiotic forms such as books, music recordings, and movies than what we retain of their EPs in our brains. For example, one can easily store all of Elvis Presley's lyrics, most of his song recordings, and many of his television interviews and films on their personal computing device's memory. Yet only a fraction of this material can typically be retained in one's mind with any fidelity. Thus, if we presume that this abiotic-versus bio-EP distribution is indicative of the global distribution of celebs' EPs, we can see how the relative weight of abiotic-EP in comparison to bio-EP is constantly increasing.

In the private sphere, where the EPs of the majority of people reside, the ratio of abiotic- to bio-EP varies widely and there are large differences in people's level of immersion in the digital world; for some commoners, the abiotic- to bio-EP ratio may be considerably greater than that depicted in the diagram. On average, though, the growth of abiotic-EP is huge, and all signs indicate that commoners' EP distribution is becoming very similar to that of celebs. The stark historical divide between celebs and commoners is disappearing, and a continuum of EPs of various sizes but with similar characteristics is forming.

Cleopatra's Fountain of Youth—EP Survival and Renaissance

Despite the rapid growth of most modern people's EPs, the underlying motivations for its creation remain mostly constant. The drive for personal, cultural, biological, and religious continuance arguably lies at the heart of most people's aspirations and actions, and has remained unchanged over the millennia. The abiotic-EP's proficiency at achieving ongoing influence in the world is a testament to its competence in this area. In particular, it has an impressive track record relating to the long-term survival of the legacies of celebs. In the previous sections, we saw that commoners' EPs are becoming increasingly similar to those of celebs. Thus, by studying the factors that led to the development, survival, and revitalization of celebs' EPs through history, we may gain useful insight into the future EPs of the majority of ordinary people.

To illustrate this point, let's take the example of Queen Cleopatra VII of Egypt (or Cleo to her friends), who lived in the last century BC, and consider how her EP evolved over the last two millennia.[5]

Figure 2 shows the schematic timeline of Cleo's EP alongside that of a commoner under her reign. The width of the timelines signifies the extent of the EP at a given time. The dimensions are shown not to scale but rather to indicate relative trends and values. During her life, Cleo's

---

[5] There is nothing exceptional about my choice of Cleopatra—I could have referred to any number of historical people to illustrate similar dynamics. Also, this discussion is not overly sensitive to the accuracy of the depiction of specific historical events, but rather is intended to give a feel for the forces that influence the persistence of such EPs.

EP is much wider than that of the commoner, signifying that she is much more influential and widely known than most of her contemporaries. However, her bodily presence and mind, represented by the internal "pipe," is similar in extent to those of her subjects. In both cases, the EP diminishes significantly after the individual's death, especially after a generation or two. That of the commoner is essentially extinguished after the last living person who knew them died. For the queen, however, there is continuity.

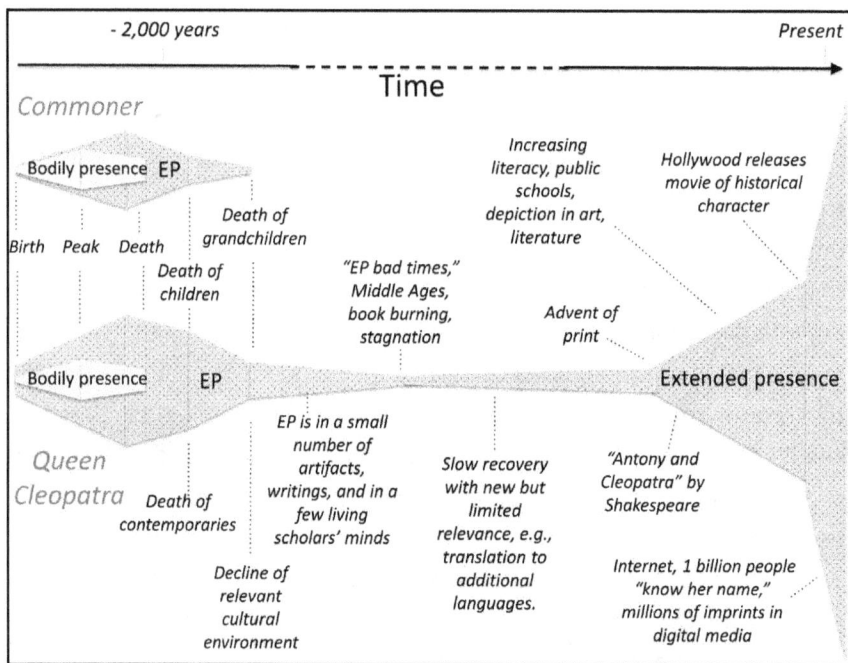

Figure 2: Historical EP of celebrity (Cleopatra) vs. commoner (schematic time line, comparative—not to scale).

Numerous inscriptions, coins, statues, and monuments depicting her were created during her reign. Some survived the centuries. Since she was well known in the Mediterranean and interacted with other regional rulers such as the Roman Caesar, her life was well documented by various historians in multiple languages. Cleo was charismatic, beautiful, sexy, and outrageous. She networked and bedded with powerful allies—rulers and generals—and employed shrewd political and publicity tactics. For instance, she portrayed herself to her subjects as the reincarnation of the Egyptian goddess Isis, thereby implying her immortality.

For many centuries after her death, few aside from scholars knew of Queen Cleopatra. At that time, literacy was limited and history was seldom taught in schools as it is today. Her legacy was carried by a few writings in various languages, and preserved in monuments and other artifacts, many of which were buried in tombs or under the rubble of ruined civilizations. Presumably, somewhere in the Middle Ages, long after the decline of the Egyptian and Classical Roman empires, Cleo's EP reached an all-time low. Then, astoundingly, in the last few centuries, it rebounded, fueled by technology and a cultural renaissance. Printing matured, literacy increased, and art flourished. Interest in Classical history spread. Shakespeare wrote a play and Massenet wrote an opera, both depicting Cleopatra. She was featured in many contemporary works of fine art. More recently, she has been portrayed in numerous movies, plays, and other manifestations of popular culture.[6]

Cleo's EP, which was semi-dormant for ages, is now thriving. Each new book, play, and movie representing her character continues to fuel this growth. Every recently uncovered coin depicting her face adds credibility to her heritage. Just as Cleo herself is alleged to have done during her lifetime, her EP is constantly reinventing itself and adapting to changing circumstances.

It is entertaining, and instructive, to imagine a contemporary of Cleo's, say, one of her maidservants, being magically transported through time to view a Hollywood portrayal of Cleopatra. What would she think while viewing her first movie, in which Elizabeth Taylor depicts her mistress? After witnessing Cleo's "rebirth" and the reenactment of iconic scenes from her queen's life, she would most probably be astonished by the power of Cleo's "ghost" to possess Taylor's body, and

---

[6] During my research I discovered that Cleopatra has even been depicted in an adult movie with much of the "plot" set in modern Egypt—the outdoor scenes complete with pyramids, camels, and tourists.

would immediately retract any doubts that she might have held about Cleo's claims to immortality.

Cleopatra's EP has achieved far more than anyone could have imagined back then. Today, the number of people who know her name is probably greater than the total world population when she lived.[7]

If, at some time during the Middle Ages when Cleo's EP was perhaps at its weakest, people had considered its prospects for continuation, what would they have thought? Looking back in history, they would have noticed a steady decline in the appearance and significance of her legacy. They would likely have surmised that her EP would eventually die out. New queens and kings were being born and creating their own fresh EPs—ones that were more tangible and relevant to the times. There was little reason to expect that Cleo's EP would undergo a subsequent renewal and then exponential expansion.

What was Cleo's EP's, or her ghost's (in the eyes of her imaginary time-transported maidservant), approach for achieving such an enduring presence and astounding revival? In retrospect, many of the contributing factors are fairly obvious. During her lifetime, she was a celebrity in the timeless sense of the word: charismatic, beautiful, influential, and eccentric. In seeking longevity, her EP seemingly pursued multiple tracks. In addition to the bio-EP in others' minds, various forms of abiotic-EP such as coins, statues, monuments, writings, and other media carried her imprint. To increase her influence, Cleo affiliated herself with contemporary Egyptian gods and other celebs, and she was multilingual and well networked at home and abroad. If the documentation by Egyptian scribes did not survive, those by the Roman historians might. Even the purported events around Cleo's demise were sensational and memorable: her defeat in battle; the suicide of her lover; and her death by snakebite to her breast, which is the kind of resonating image that, now as then, is envisioned by people once, and then forever remembered.

---

[7] A widely quoted estimate of the world population at the start of the Christian era is 200 million people. Even if this were double, the assertion would probably still remain valid.

Cleo's Style—Adaptation and Relevance

Cleopatra's EP has been developing and adapting to the world for the last two thousand years; she has kept in step with the times. It is misleading to think of Cleo primarily as a page in a history book. Her bodily lifetime is only a small aspect of the *current* Cleopatra. Indeed, history is not an assortment of frozen facts of what happened back then. Rather, it is a theory of development that expounds, in current terminology, how things came to be as they are now. If a coin depicting Cleo is uncovered today by a team of workers using a backhoe while laying fiber-optic cables, and the location of its discovery is recorded using location-determining technology enabled by satellites, then all these things and more are part of Cleo's unfolding history.

It is the current perception of Cleo's entirety that is most relevant to her perpetuation. This perception, however, is influenced primarily by events and developments that occurred long after she lived. Much of what is known about her is derived from the writings of Roman and other historians, some of which were composed decades or more after her death. These writers were depicting *their* perception of Cleopatra from their frames of reference. Their descriptions were already deviating significantly from anything that could naively be conceived as "pure historical fact." The primary evidence for this assertion is that even back then, there were significant differences among the various accounts. The same, of course, is true today for news outlets describing the events of yesterday.

The key for the EP's survival and expansion is its ability to develop and adapt, and to remain relevant and valuable in changing environments. The notion of pure fidelity, or the retention of things exactly the way they were, plays a much lesser part in the EP's continuation. Coins depicting Cleo's face may remain almost unchanged over the centuries, but the more expressive depictions of her character—like those we experience in books and movies—are constantly changing and conforming to the times. It is these latter-day representations that are primarily responsible for the rapid growth of Cleo's EP.

Cleo has often been portrayed as a symbol of beauty. A look at how her physical attributes, as interpreted by contemporary cultures, have changed over the ages illustrates the adaptive nature of her EP. For example, Cleo's nose has been shrinking for the last two thousand years. During her lifetime, a prominent nose was considered attractive and especially befitting a ruler. This distinctive element of her profile was described in writing and is also evident on coins and sculptures from that period. Observing Cleo in recent times, as she is depicted in movies and theaters, we clearly see that her nose has adapted to modern tastes. Similarly, the shape of her figure has fluctuated according to the fashions of the eras. In the seventeenth and eighteenth centuries she gained weight, as is apparent from her portrayal in a number of semi-nude paintings. Later in the twentieth century she became slimmer again, in line with Hollywood's ever-changing sensitivities to the ideals of body form.

For living beings, survival requires movement, and movement means change. Trying to keep things just as they always were is a futile exercise best exemplified by fossils and mummies. These have their significance and can preserve a likeness of appearance for a long time. Nonetheless, they are dead and will always be very partial representations of the originals. Similarly, archives can decrease rates of decomposition of their contents, but they are also subject to inevitable decline.

By contrast, a living organism like the silverfish insect, or fish-moth, which incidentally is also adept at expediting the disintegration of books, has mostly retained its form for millions of years. A modern silverfish is more similar to a living silverfish from a hundred thousand years ago, than either is to a silverfish fossil.[8] Archives and fossils are

---

[8] What does it mean to say that a silverfish from a hundred thousand years back is similar to a modern silverfish, and less like a fossil that was formed from a silverfish that lived a hundred thousand years ago (say the former silverfish's sibling)? While we could easily get into a drawn-out discussion on this subject alone, the basic idea is that in characteristics such as material composition, metabolism, behavior, and perhaps even the ability to interbreed, the time-traveling ancient silverfish and the modern individual would be quite compatible. None of these comparisons would be valid for the fossil.

static remnants of the past. Living and adaptable beings, thriving in present-day environments, are things of the future.

My point here is that a developing and long-lived EP behaves more like a living organism than an archive or a fossil. Pure fidelity, or sameness, may be compromised, however a core character and intrinsic relevance can endure. Indeed, often what we call "loss of fidelity" is not a loss at all but rather the apparatus of advancement into the future.

The EP adapts and becomes animated to remain relevant. Cleo's EP moved and transformed, thereby surviving EP droughts like the Middle Ages, and eventually flourished when the environment improved and new, hospitable niches such as digital media emerged. This is not a dry history of old books on dusty shelves but a living, pulsating, and growing EP—an entity that is relevant and valuable today and well poised to prosper in the future.

Cleo's Envy and the Girl Next Door— Perpetuity for Everyone

A number of thinkers in recent years have compared the absolute affluence of an ancient monarch to that of a socially and economically average person in the developed world. Indeed, in many fundamental ways the modern person is much wealthier than the kings or queens of old. The modern person eats a larger variety of food, drinks safer water, and is warmed by the flick of a switch. As a parent, she does not have to watch her child die from an infectious disease that is now easily cured by antibiotics. She can drive a car in safety at speeds unimaginable for the queen of ancient times. The modern person enjoys anesthetics for dental treatment and childbirth, and typically lives much longer. We can go on and on with similar comparisons.

But most astounding, perhaps, is the vastly greater *affluence of information* that is now available to almost everyone in the developed world. This has progressed far beyond the wildest imaginations of the sages of old. A literate pauper can leave his bags at the entrance of a public library and freely access the Internet or the library's collection of books. He can then discover what is happening on the other side of the world, glimpse the bottom of the deepest seas, view the earth from space, and even see images from Mars. He can learn the solution to almost every question that has ever been answered or sit back and tour the art collections of many of the world's most revered museums.

Some may argue that the benefits of this generalized affluence are mitigated by downsides, such as environmental degradation, the complexity of daily life, and so forth. However, people have clearly indicated their preference. Throughout history the vast majority have gratefully adopted these advances. Moreover, virtually no one, despite the real possibility of doing so, turns their back on the comforts and safeties of our modern lifestyle.

Here, my intention is not to either celebrate or bemoan the profound changes brought on by modern technology but rather to extrapolate these trends to envision the implications of modernity for the EP. In a number of senses, Cleopatra has achieved *perpetuity*. Her EP has not only survived for two millennia but it has actually grown and flourished. In fact, with the advent of the information age, and barring the demise of civilization as a whole, it is hard to conceive how her EP could *not* continue indefinitely.

I would argue that a comparable type of perpetuity is now within the reach of ordinary persons. Moreover, the EPs of most modern individuals can surpass those of ancient monarchs by almost all possible measures. The girl-next-door's EP can achieve the longevity of Cleo's EP, but with a much richer and profoundly more personalized reflection and extension of the girl's being.

Figure 3 shows how a commoner's and a celebrity's EPs will develop in very similar ways in the future. The total "width," or extent, of the celeb's EP is obviously greater, yet the general trends and potential longevity are equivalent. Both instances exhibit continuous future growth.

Seeds, though sometimes dormant for centuries, can germinate and flourish when the environment becomes conducive to their growth. As Cleo's past so aptly demonstrates, the same is true for the extended presence. While many parts of an individual's EP are relevant and useful today, other parts can be regarded as spores awaiting the emergence of favorable conditions. Created properly and maintained well, robust and adaptable EPs have the potential to deliver a host of further applications—functions that are presently unimaginable, but will almost certainly exist in the future.

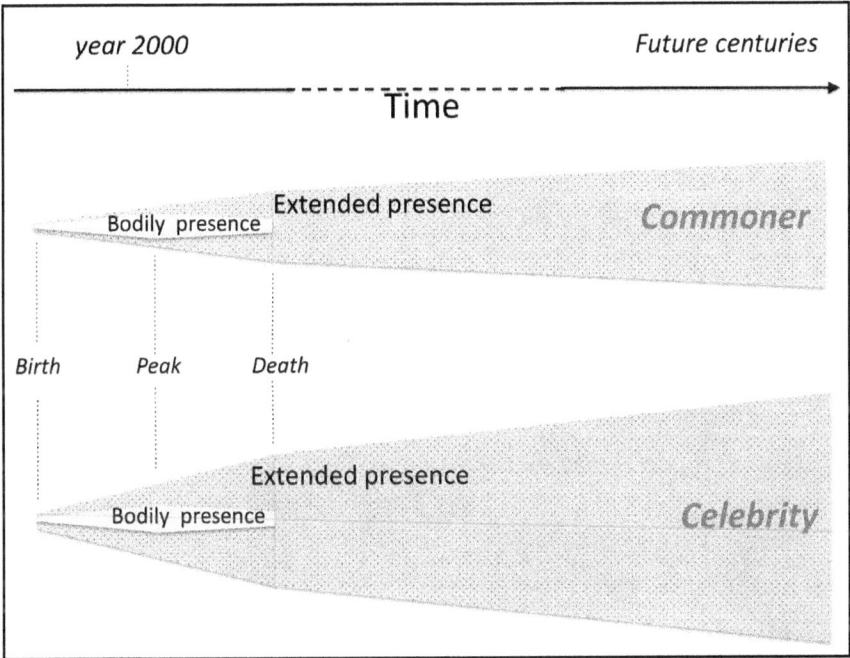

Figure 3: Future EP of celebrity vs. commoner (schematic time line, comparative—not to scale).

CHAPTER 2

EXPRESSIONS OF EP VOLITION

When we observe celebrities' EPs, it often appears that they exist with varying levels of independence from the originating people. It's not always clear who's at the helm—the EP or the physical being. Let's look at some examples to illustrate this notion of the EP possessing apparently semi-autonomous and self-interested traits, and explore whether, in this area too, the character of celebs' EPs indicates likely outcomes for those of the broader population.

Fame, Death, and Heroes of the EP

History is replete with instances of celebrities amplifying their fame at great personal expense, in particular to the detriment of their physical and mental health. In many cases, the circumstances of these celebs' fame seem closely linked with the causes of their death. Moreover, their demise does not necessarily diminish their fame over the long-term. On the contrary, a stunning death often captures the public imagination and perpetuates the celeb's fame far more than a gradual decline into old age. One needs only to recall the circumstances of the death of one's favorite rock star, president, princess, villain, or other celebrity, and consider whether, and how, their death and fame are related.

How can we understand such tragic outcomes and self-destructive behaviors? Is the EP a person's device entirely, under their control and for their material benefit, or is it rather the EP that gains agency, exploiting the misfortune as well as the fortune of the physical person in order to thrive?

The concept of martyrdom is an especially striking example of the significance of an esteemed and memorable death. If we look closely, we see that it also illustrates that accepted ideas of what is real and tangible are often more nuanced than we think. In particular, it implies the existence of an influential—and even a self-interested and manipulative—EP, which provides a new framework for viewing certain human behaviors.

Let's examine this more closely. What is a martyr? Has anyone ever observed or touched one? If not, how do we know they're real? If we say someone is or was a martyr, whom do we mean by the "someone"? For simplicity, let's consider a case in which the person dies instantaneously. Clearly, the person cannot be a martyr before their death. Further, we do not think of the dead body as the martyr. Thus, a martyr is neither a living nor a dead person in a bodily respect.[9] Let's take this a little further. Can a martyr exist outside of the perceptions of others? Maybe in theory, but such martyrs are, not surprisingly, unknown!

So what have we established so far? First, a martyr is not a person in any bodily sense. Second, in practice, a martyr can only be created in other people's brains. From there, it can be expanded and grown via media such as books, movies, monuments, street names, etc. Does this mean that a martyr is initially just a notion in our minds—a figment of our imaginations? Although we imagine martyrs to be very real, when we try to examine them in any detail, they really seem to be quite imaginary.

---

[9] By contrast, a concept such as "hero" does not have these characteristics. A hero is an attribute of a living person. We say that someone *is* a hero or, if deceased, that he or she *was* a hero.

Most of us are familiar with at least some of the explanations and motivations for martyrdom as depicted in history and in the daily news. These vary significantly depending on the culture, period, and related religious interpretations. Yet, none of these explanations resolves the inconsistencies we've noted. From an EP perspective, however, the concept is much more coherent and tangible. Depending on the prevailing cultural norms, in a characteristic type of situation, a series of physical actions, culminating in a person's bodily death, leads to the formation of a distinctive type of EP in other people's minds, which we call a martyr. Typically, this involves the almost immediate conversion of a single bodily presence into multiple EP instances, or units. Under certain circumstances, this transformation can result in astounding conversion factors—one life "exchanged" for millions of long-lived EP imprints.

Mohamed Bouazizi, for example, the Tunisian street vendor who is credited with triggering the Tunisian Revolution in 2011, is now known in his country and internationally as a martyr. He ignited the revolution quite literally through self-immolation. Bouazizi's EP has spread extensively. Numerous articles document his life in many kinds of media, and he is featured in new encyclopedias. His bio-EP is carried by millions of people. He[10] was awarded the Sakharov Prize and named "Person of the Year" for 2011 by the UK newspaper, *The Times*. Tunisia has issued a postage stamp bearing his image. His EP, once unremarkable, is now assured a substantial and long-lived existence into posterity. Intriguingly, there are no indications that Bouazizi was thinking about national affairs or revolutions during the events that led to his death. For him, the action was chiefly a practical and personal matter. From news accounts, it appears that he was primarily concerned with the return of his street-vending weighing-scales, which had previously been confiscated by municipal officials, and redress for the humiliation he had

---

[10] I repeatedly use the term "he" in this paragraph but clearly this cannot be in a "he the person" sense. "He the martyr" as a purely EP construct is closer to what I mean. Later in the book, we will discover a more consistent terminology for describing such situations.

suffered at their hands. His EP, however, apparently had other plans in mind.

Here we see that, even though the longevity of one's bodily presence typically also promotes the growth of their EP, under certain conditions, a shortened life culminating in martyrdom can be an effective EP growth strategy. For certain individuals, this may be the fast track or perhaps the only track to EP riches, or what we more commonly call fame.

The consistent view of martyrdom in the EP domain, and, by contrast, its nebulosity in the bodily domain, is a consequence of martyrdom being primarily an EP phenomenon. From an EP perspective, martyrdom is also concrete in that it can be measured in a number of ways. For instance, we can test whether an image of a specific martyr exists in a person's mind by showing them a collection of similar photographs with only one depicting the martyr. If the person can correctly identify the martyr's picture when given his name, then we know that they carry this aspect of the martyr's bio-EP in their brain. Recently, more direct methods such as various forms of neuroimaging are being used to infer the existence of or lack of specific information in one's brain. The martyr's abiotic-EP is also quantifiable through parameters such as the number and extent of references in news articles, books, movies, and street names.

Additional examples abound, in literature as in life. In an iconic scene in Tolstoy's *War and Peace,* before Prince Andrei leaves for the war, his father cautions him that shame is worse than death. Dying relates to the body, but shame relates to the person's and their family's EPs. In his father's eyes, the status of Andrei's EP, or reputation, is more important than the well-being of his body. This scene is from a work of fiction, yet it depicts a very real mind-set.

Greek Spartan mothers are said to have similarly instructed their sons before battle: to return on their shields (dead) rather than without their shields (humiliated). Analogous attitudes incite duels and other tournaments of honor, in which the perceived gain is typically not of material value but, instead, of repute, or EP quality.

Likewise, soldiers are sometimes posthumously awarded Medals of Honor. But who actually receives the award? Usually we do not say that the family is being awarded the honor. Rather, we insist that it is the individual who is receiving the citation. Clearly, this is not the person in any bodily sense; in modern warfare especially, the body often completely disappears. Rather, it is the non-bodily part of the person—their EP—that is actually being awarded the medal. If the medal and other forms of recognition contribute to the EP's longevity and dissemination, then it is consistent to state that the soldier's bodily death benefits the soldier's EP.

In general, the prospect of recognition, meaning EP expansion, is one of the central motives for bravery. From Purple Hearts, Green Berets, Silver Stars, and black belts, to kill markings on fighter aircraft, campaign ribbons, and flight wings—this list goes on and on. If voluntary soldiers knew in advance that there was absolutely no possibility of anyone at all—person or deity—knowing about their feats, would they still so willingly endanger their lives? Under such universal anonymity, an act of bravery resulting in their death would be a very bad EP strategy.

These examples imply that, in certain circumstances and in self-interest, the EP may have a nefarious influence on people's physical lives.[11] I am not suggesting, however, that this is the EP's primary strategy. The majority of us are not dying as heroes or martyrs, or gaining eternal notoriety by assassinating presidents. Yet, this mind-set shows how certain extraordinary processes can be more coherently understood from an EP perspective. Let's now consider some gentler and more common demonstrations of EP dynamics.

---

[11] In regard to the EP's possibly selfish and detrimental effects on the bodily presence, we may be tempted to equate certain EP manifestations to forms of parasitism. Over time, however, it will be become clear that a mutually obligatory symbiosis will be a more fitting description of the bodily presence and EP relationship. Later, I will propose that the notion of a conceptual unity formed from a physical person and their EP removes any remaining contradictions.

Reputation on Earth as in Heaven

In antiquity a "great man" may have been physically superior to his peers. In more recent times, however, this term rarely pertains to one's physical attributes. Indeed, it primarily relates to the quality of their EP. Generally speaking, a great man or woman is one with a positive, widespread, diversified, and enduring EP.

Similarly, our reputations, or the status of our EPs, are of immense consequence to most of us. One's good name can have great practical benefit. It may translate into a good job, a good spouse, good food, and good friends. But, it frequently seems that a lot more than its direct usefulness is at play; the EP's good standing often seems to be an end in itself.

EPs that are primarily evaluative, or qualitative, exist in many forms, the most ancient of which are opinions in other persons' minds or feathers in warriors' headdresses. These EP expressions appear traditionally as certificates of merit, titles, medals, educational report cards, professional rankings, credit histories, and the like. More recently, our rating metrics on social media—the number of views, followers, connections, and "likes" our digital presence garners—are becoming increasingly important for our private and professional lives.

For celebrities like politicians and entertainers, the way in which others perceive them is of almost existential consequence. Yet, the opinion of others does not necessarily correlate with the reality of "the person" in the classical sense. In fact, others often believe things about celebs that are very different from how they really are or how they would appear in direct interpersonal encounters. Since most individuals rarely meet their favorite actor or politician, their impression of the celeb is generated almost exclusively from synthetic media. Thus, for most people the celeb's abiotic-EP is essentially equivalent to the person.

Until fairly recently, the bulk of most people's interactions with others was face to face. But today many of us increasingly communicate via social media and other modern methods. So, for us too, the characters we project via our abiotic-EPs may deviate considerably from those apparent in direct physical interactions. Like celebs, our abiotic-EPs are progressively becoming the primary representations and conveyers of our personalities. Today more than ever before, a growing number of the people we interact with—our acquaintances, friends, and colleagues—are rarely, if ever, physically adjacent to us. For these people, the collections of our abiotic-EPs are, arguably, *who we are*. If it is true that our abiotic-EPs are increasingly who we are, then clearly, many of our ambitions and the privileges that we have traditionally reserved for ourselves, as people in the classical, bodily sense, will extend to our EPs as well. In particular, our reputations will be indistinguishable from those of our abiotic-EPs.

In many respects, the concept of the EP's good standing as an intrinsic value is suggestive of the religious notion of an enduring individual reputation as measured by the degree of one's compliance with the directives of their belief system. In many religions, this status, frequently referred to as the condition of one's soul, is viewed as the primary factor in determining one's fate in posterity. Traditionally, ensuring one's religious good standing has also been a chief motivating factor and practical directive in one's daily life. In the context of this discussion, I suggest that the prestige of our abiotic-EPs will attain similar prominence and become a primary motivator of our actions in the "physical" world. Already, many people serve their abiotic-EP by conforming their behavior to fulfill its apparent needs. For instance, individuals often perform an action expressly to record and post it on their social network. Likewise, professionals sometimes make a decision that is primarily determined by how it will affect their resume.

We record and perpetuate even the most Hedonistic and live-for-the-moment of our activities. In effect, we are systematically creating memories and reputations. The writing and photography and filming are not merely documenting events impartially from the outside but are evermore becoming an integral part of the events. From weddings to wars, the recording permeates into and transforms the occasions.

The status of our abiotic-EPs is increasingly influencing our current lives, and over time, like Cleopatra and Shakespeare, it will be perceived as determining our standing in perpetuity.

Reunions, Ceremonies, and Other EP Refreshments

A *reunion*, the temporary coming-together of a previously disbanded social group, is another example of seemingly self-interested EP behavior. What is the motivation for meeting with others with whom we once spent significant time, but who are currently not a regular part of our lives and will likely not be in the future? Why do we care to participate in these gatherings or sometimes feel angst if we miss out?

Let's consider this behavior from an EP standpoint. Because of the substantial and intense time we once spent together, perhaps during very formative parts of our lives—high school, college, military service, etc.—our EPs have a high level of commonality. Besides our family and close friends, these people are often among the strongest carriers of our EPs. Typically, some of our high school or army buddies will forever remain in our memories, and we in theirs.

Thus, we can think of the "reunion urge" as our EPs encouraging us to make the effort to meet up. Because our EPs are already firmly rooted in these people, getting together for a few hours is a very efficient way to refresh and update our EPs in their minds and vice versa. This is in contrast to meeting them each separately, where the effort spent would be much greater without a corresponding increase in EP benefit.

From a long-term EP perspective, this EP gain would also be far better than the potential new bio-EP we would achieve in a one-time gathering with strangers (say of the same duration as the reunion) where, most likely, no one would remember us five years hence. From the EP's standpoint, these considerations are analogous to the superior economy of investing to refurbish an old but well-crafted building, as opposed to expending a similar effort to erect a new but flimsy construction.

In addition to the actual reunion meeting, which directly reinvigorates the bio-EP in the participants' minds, modern media enhances the abiotic record of the event with images, video, and social media, rendering the EP reinforcement exercise even more effective.

In short, although it may be somewhat difficult to identify any practical reason for occasionally reuniting with old acquaintances, in EP terms it is perfectly rational. Consistent with the notion of a sometimes seemingly egocentric EP, it is the EP, in a sense, that drags the reluctant body along to the reunion. The body may have little interest in this game; it would rather stay home and get a good night's sleep.[12]

In this light, we can reconsider familiar notions such as nostalgia and reminiscence. Nostalgia is perhaps the pain of an aging and waning EP, while reminiscence is one of its remedies. In the same way that our ability to feel bodily pain is essential for our physical survival, it seems that our capacity for emotional suffering may be instrumental in the survival of our EP.

A reunion is but one example of a ceremony. In general, many ceremonies and other rituals are occasions where very little constructive activity seems to be happening. From an EP perspective, however, ceremonies are very productive events where existing EPs are being rejuvenated or merged, as in memorial services or weddings, or new EPs are being established, as in infant-baptisms or inaugurations.

A wedding, for instance, can be viewed from an EP perspective as a process whereby transitory bodily gratifications, such as food, drink, and dance, are exchanged for long-lasting EP gains.

Cleo's Latest Affair—Cooperating EPs

One trait of autonomous beings is that they can choose to cooperate with each other for mutual gain. Let's look at whether and how EPs exhibit such behavior.

Here it's instructive to revisit Cleopatra, this time as portrayed by the actress Elizabeth Taylor (Liz) in one of the costliest movies of all times,

---

[12] Notably, even in the absence of physical meetings, in social networks, a significant part of the more mature users' activities is focused on reconnecting with old acquaintances. Most probably, the younger generations will never be in this position; that is, they will never need to *reconnect* since they will easily be able to retain a basic, life-long connectivity with their friends and colleagues.

Cleopatra (1963). From our current perspective, we can think of the movie as a cooperative venture between Liz's and Cleo's EPs—an affair in which each benefited from the other's reputation.[13]

Before acting in the movie, Liz learned about Cleo. She probably read some books, saw some plays and previous movies, and then used her body and mind to communicate Cleo's character.[14] In other words, Liz portrayed a merging of Cleo's and her EPs. Or in a sense, Liz allowed Cleo to "possess" her person during the filming. They brought together their complementary types of fame: Cleo, as someone ancient and well established, and Liz, as someone contemporary and original. Each benefited from what the other had to offer: Liz gained an aura of timelessness, which Cleo had refined over the millennia. Cleo gained some of the modernity and freshness that Liz had in abundance. Consequently, Liz's and Cleo's EPs each came away from the liaison fortified and more influential.

Clearly Liz the physical person was also part of this transaction. She was bodily present at the filming, and a good many material things—money, property, travel, and more—came her way. However, this material aspect was short-lived and ultimately comprised only a small and transient part of the larger, ongoing story. The affair between these two powerful EPs was, and still is, the more significant and long-lasting part of the alliance. Once the filming was complete, a dynamic entity had been created that was mostly independent of the physical existence of the EPs' originators' bodies.

The creation of the movie spurred the growth of their respective EPs in a myriad of ways. Their EPs were replicated in various abiotic media and then spread into millions of people's minds. A multitude of

---

[13] I use this example not to provide an accurate description of the events surrounding the 1963 movie, *Cleopatra*. My purpose is to impart a sense of the processes involved, and to view such dynamics in the context of the EP.

[14] While phrases such as "She used her body and mind" are not altogether unusual in our language, in the current context, they may be a bit perplexing to some readers. We may ask: Who is the "she" if not the "body and mind"? I attempt to reconcile this quandary later in the book with the idea of a union of a person and the collection of their EP constituting the "she."

offshoots ensued. Authors wrote reviews and books, and many others spoke and dreamed of Liz and Cleo. The effects continued to multiply. Having seen the film, some were inspired to read historical accounts of Cleo. A school curriculum committee might have decided to include Cleo in the subsequent year's syllabus. Through these actions, Cleo's EP prospered.

Then, in subsequent years, history instructors, while teaching Cleo's story, recommended the movie to their students. These students were thereby introduced to Liz's EP. Having enjoyed her performance, these students went on to see her other movies and read her biography. They may have become fans. In this way, Liz's EP flourished and grew.

Today, with the actress and the queen long gone in the bodily sense, their EPs still have the smooth skin and seductive features of the past. In fact, they have more than that—they are alive, growing, and replicating. I'm writing about them. At this moment, you are reading about them, and yes, bits of them have just crept into your mind and may remain there for quite some time.

The dynamics of EP collaborations resulting in mutual benefit may not always appear so harmonious. Consider the assassination of president John F. Kennedy, apparently by Lee Harvey Oswald, who was also killed a few days later. One can reasonably argue that while we will never know how Kennedy's life would have played out had he lived on, the circumstances of his assassination and its aftermath have greatly increased his fame.

Additionally, Oswald, who would probably have remained an unknown figure if he had not been named the assassin, thereby gained immediate, widespread, and lasting notoriety. If these contentions are true, then it is consistent to conclude that although Kennedy's and Oswald's bodily existences were starkly conflicting and abruptly ended in 1963, in a sense, their EPs were congruent and greatly benefited as a result of their interaction.

In more modern contexts, we see our EPs colluding in a multitude of new ways. This is perhaps most prominent in social media where almost all actions can be understood from the perspective of cooperating or sometimes competing EPs.

From Shadow to Future Self

The examples presented here imply that to fully grasp the EP's nature we eventually need to abandon the idea that the EP is purely and entirely a byproduct of a person. That is, the EP is not merely a static object or a compendium of exterior entities affiliated with the person. In fact, in many circumstances, it appears to be a fundamental entity with an existence of its own—perhaps even possessing a degree of self-volition.

The fact that the EP originates from people does not contradict this notion. People themselves have originators—their parents, but they eventually become independent. Similarly, we will find that although the EP may be initiated and nurtured by an individual's existence, it is subsequently able to attain various levels of autonomy. Moreover, in certain circumstances, the EP may exhibit what appears to be self-interested behavior, and become influential in ways that are no longer plainly linked to the person from which it originated.

At first, endowing the EP with a degree of autonomy may seem strange or objectionable. As an individual, we typically perceive our essence—our self—to be fully contained within us. We deem all else associated with our personhood as subordinate. We measure our EP constituents by their utility to us or view them as byproducts of our physical existence.

For example, while we usually consider an artistic creation to be a *product* or *expression* of the self, we seldom see it as a *component* or *extension* of the self. We are even less disposed to regard it as an intrinsic entity with characteristics akin to the self, or one that has a capacity for independent existence. I contend, however, that as we progress in our exploration of the EP, we will discover numerous reasons to replace this mind-set with one that is more compatible with emergent reality—we will find that the personal EP sometimes seems to take on a life of its own.

Part II

EMERGENCE

CHAPTER 3

HUMEM-INDIVIDUALS

In the previous chapters, we reviewed the history of our extended presence (EP) and the huge recent growth of our abiotic-EPs, particularly in the form of digital data. We saw some ways in which celebs' EPs have developed, adapted, and endured for centuries, and how in current times, the EPs of celebs and ordinary persons are becoming ever more alike.

We also noticed that our EPs have some complex and subtle interactions with our bodily presences. In certain settings, EPs can be seen to influence the actions of physical people no less than the actions of people influence the formation of EPs.

We now continue our exploration, begin to venture into uncharted territories, and prepare for our first glimpses of the offspring of these developments. In this chapter, we'll see some ways in which the EPs of individuals are beginning to consolidate into entities with distinct characters and abilities.

Current Fragmentations and
Early Aggregations

Even substantial and long-lived EPs have been widely dispersed throughout much of history. For example, Cleopatra's EP resides on a broad variety of media such as coins, parchments, stone statues, books, electronic media, T-shirts, and so on. It is geographically scattered—

owned by various individuals and institutions, or still buried under ancient ruins. Consequently, the connections among these various parts of her EP have been severely degraded, and in some cases completely lost, since their creation.

Today, our own abiotic-EPs are fragmented and similarly dispersed over many kinds of media and applications. Moreover, current EPs are often transient and fragile. As digital technology develops and improves, older EP constituents are frequently abandoned and ultimately lost. Due to constantly changing formats, the high sensitivity and rapid degradation of modern data storage media, and a host of other digital-data maintenance difficulties, modern forms of data often have much shorter life spans than traditional types of records like clay tablets, paper books, and oil paintings.

Many people already perceive their data as an extension of their mind and memory. Seen in this light, fragmented personal data is akin to a divided mind or a dispersed identity. This is evidenced in a myriad of ways: emotionally, for example, by the missing synchronization between one's old diary and contemporaneous photographs (*Was that picture taken of me before or after my graduation?*), or practically, as the forgotten correlation between one's physical location at a given time and a record of a simultaneous in-store credit card transaction (*Did I really make that purchase?*).

Another simple but representative case of fragmentation is when correspondence relating to a single topic or event occurs via separate media, such as digital voice communications and email. Each of these channels conveys complementary and interconnected aspects of the event. However, because they are disconnected, a separate recording, or digital memory, of each cannot preserve the completeness of the original interaction. A question may be asked via one channel and answered via the other. As long as the channels are detached, an orphan question (one without an answer) would remain in one data body, while an orphan answer (one without a question) would remain in the other. In this case, if the record in each channel contained an identifier linking it to the event and a time stamp, a later reconstruction of the question and answer communication would perhaps be possible.

However, in more complex interactions, especially those in which multiple forms of media and many applications are involved, the relevancy contained in the interconnectivity among the various channels is irretrievably lost. As a result, with respect to the collection of the recordings, or memories, of these channels, the sum of the parts is considerably less than the original whole.

Nevertheless, abiotic-EP dispersal is not universal. Historically, and also more recently, there have been various types of personal EP aggregations. A traditional example of EP consolidation is a book or a collection of books by or relating to a specific person. In particular, memoirs or biographies can be seen as deliberate efforts to consolidate a person's fragmentary EP.

A collection of an artist's works is another classic example of a deliberate bringing together of a personal EP. Typically, the artist may sell pieces of their work over a long period to various customers, resulting in a wide dispersal. But, if the artist subsequently becomes sufficiently appreciated and famous, substantial worth may be placed on a temporary or more lasting gathering of their works. This is the value of collections.

Now, collections can sometimes be regarded as simply more of something good located in one place. This is certainly part of their value. Yet, almost without exception, these assemblies result in much more than just a sum of their separate constituents. A substance of interrelationships is reestablished. *Bodies* of work are created. In a collection of paintings, for example, an exhibition can depict the development of the artist's creativity over time; a story emerges. For many observers, viewing and understanding one work can enhance their appreciation of another. Other instances of inter-relevance abound. For example, a book depicting a collection of an artist's paintings, together with commentary and supplementary contextual information, can achieve similar results.

Today we see an ever-increasing proliferation of personal EP aggregations, from photo albums and personal or familial websites, to user channels on video-sharing websites, social media accounts, and so on. Much of their significance lies in their interconnectivities and mutual relevancies—that which glues them together. Moreover, as exemplified by social media, similar relevancies exist at other levels such as in the relationships among the EPs of different individuals and groups.

Taking Shape—Coalescence and Identity

Now let's focus on the aggregate of all of the components of an individual person's EP. Can we view this collection as a conceptual unit—a whole possessing form and identity?

One's bio-EP in other people's minds is clearly beyond one's direct control. One's abiotic-EP is also fragmented and dispersed over many media and multiple jurisdictions. Much of it is in the public domain or is held by corporations and governments and is therefore not owned or administered by its originator, who is often oblivious to its existence.

In other cases, one's abiotic-EP is more consolidated, with its originator maintaining partial levels of access and control. For example, one may have a publicly accessible medium such as a personal website or an account in a social media application. Although the access may be partially or wholly public, and one may not even formally own the material, one may still possess some level of control by being able to designate much of the content.

Other parts of one's abiotic-EP are much more private and tightly controlled. One may even legally own them. Examples are diaries and printed photographs or the mass of private electronic data in one's personal storage devices or online repositories.

In modeling the distribution of one's entire personal EP, we can envision a concentrically layered sphere, well defined and dense in its center, and gradually becoming fuzzier and less cohesive further out. The distinct core represents the private and controlled parts of one's EP, much of which may be concealed from entities outside the sphere. As we start to move outward, we see layers representing parts of the EP that are still clearly affiliated with the core but with decreasing levels of control and privacy—becoming ever more visible and accessible to others outside the sphere. Even further out, we encounter more detached and nebulous EP fragments with ever-diminishing affiliation with the center. As the distance from the core increases even more, we find bits of EP whose affiliations are unclear. In some cases, these "floating" bits of EP may be associated with more than one individual, or have no clear association at all.[15]

Furthermore, the abiotic-EP is not just a collection of inert objects—it exhibits many kinds of dynamic behavior. We can loosely define the *EP dynamics* as all the things that the EP does or can do, or that are done or can be done to it. Historically, human intermediaries have facilitated most of these EP actions and abilities. Increasingly, machines, especially networks of computers, are empowering these dynamics. As it grows, the compendium of EP dynamics associated with a specific individual's EP also acquires a unique quality.

When considering the totality of one's EP and its dynamics, even though its composition and boundaries are not precisely defined, we can gradually start noticing a body with a distinctive core and character. We can also begin recognizing an identity, typically correlating with that of its originator. As is customary for anything new that has come to stay, we need to give it a name: let's call the coalescence of one's EP together with its dynamics, a *humem*.

When You See One, You'll Know One— Recognizing Humems

But what exactly is a humem? Where does the humem start and where does it end? There are no simple answers to these questions because there are, and always will be, substantial variations in humem expressions. The difficulty in defining a humem is similar to that which we encounter when we try to define a person. In fact, these concepts are so closely intertwined that it makes sense to briefly examine what we

---

[15] A classic example of substantial but unaffiliated EP is a person's contribution to an online crowd-sourced encyclopedia. By design, this kind of EP is not associated with a specific person. The same is true for many other collective endeavors. While the overall design of a large building, for example, may be clearly affiliated with a certain architect, the EP of one of the many bricklayers is typically much harder to assign to its originator.

mean by a person. This review of personhood will help clarify why the notion of a humem is inherently variable, developing and fluctuating in time and context. Moreover, we will find that even though we may be unable to specify a universally valid definition of a humem, once we become familiar with these entities, we will easily and intuitively recognize them.

So, how do we define a person or personhood? This may initially seem easy. However, the more you try to pin it down, the less obvious it becomes. Even the relevant professionals, such as philosophers and legislators, are not very successful in this area. Most definitions are valid for practical and localized uses only. Biological approaches, such a defining people as "bipedal primates," are not very helpful when we look at the bigger picture. While the biological body does of course constitute a core component, a person is more fully understood as a broader system of more or less tangible parts, along with connections to other systems such as clan, society, state, and law. That is to say, for most purposes, it is more useful to comprehend a person in an anthropological context.

Let's think for a moment about the adjectives commonly used in conjunction with the word *person*. In everyday speech, we use combinations such as "tall person," "healthy person," and "small person." These clearly pertain to one's bodily attributes. But, terms such as "good person," "wealthy person," "influential person," and countless others, have little to do with one's body.

Furthermore, each of these phrases has a wide range of connotations depending on the context. Take the "wealthy" attribute, for example. Depending on the place and time, its measure could vary from the localized and tangible, like the number of cowrie shells or chickens one owns, to the highly abstract, with numbers described by complex rules and fluctuations, manipulated and expressed via computers, representing a stock-ownership based affluence. Astoundingly intricate cultural, legal, and mathematical systems determine the dynamics of some of the advanced types of ownership and wealth; very few people really understand the complexities of these various elements in any depth. Yet, amazingly, virtually no one—not even a child—bats an eyelid when told that someone is wealthy. We all, somehow, know what it means. Moreover, as abstract and virtual as it actually is, most people think of this kind of wealth as a very real thing.

A similar intuitive mind-set will help to crystallize our understanding of humem nature. For instance, we do not need to know every intricate detail of the formation and composition of persons in order to comprehend what they are and their places in the world. In particular, we know from direct experience that we needn't understand these things about ourselves to be people and to live our lives. Similarly, even without being able to fully decipher the complexities of the formation of humems or their entire makeup, we'll still be able to intuitively perceive their nature and behavior. Once we become familiar with them, it will be hard to imagine not having known them. Subsequently, despite their variability of form, when you see one, you'll know one.

Still, to facilitate discussion, a rudimentary delineation may be better than none. So let's slightly amend our previous description of a humem to the following working definition:

A humem is the product of the coalescence of an individual's EP, and the capabilities of the resulting whole.

As we progress in our study of humems, we may find that this is an imprecise description of certain humem manifestations. It is, however, mostly compatible with the leading example that we discuss in this book, namely, the humem formed from a specific person's abiotic-EP.

As we will gradually discover, humems are similar to people in many senses but without the "usual" bodies. As we come to understand them better, we will see that they also have tangible parts. We can call the collection of all the physical parts affiliated with a specific humem the *humem body*. By means of this body, the humem is able to affect and interact with the material world. Conjuring the picture of the EP sphere presented previously, we can view the humem body[16] as the merger of

---

[16] Although the context should mostly elucidate the connotation in the reuse of familiar terms, I will make extensive use of qualifiers such as "person" and "humem" in terms such as "person body" versus "humem body," or "people-state" versus "humem-state." This is intended to counter ambiguity in the many situations in which it could creep in.

the tangible parts of one's abiotic-EP—a core with tight cohesion and a periphery with fuzzy boundaries. (As such, humems apparently lack a counterpart to the clearly defined boundary of a human body—its skin.)

The EP dynamics—the actions performed on or by the EP, and the EP's inherent abilities—are also integrated and merged as the EP coalesces into a humem. Thereby, humems come to possess degrees of vitality and agency.[17]

Notably, prevailing views of personhood generally include non-bodily aspects like one's influences on society, which in the current context are also parts of one's EP. Thus, the notions of *personhood* and *humemhood* overlap. Nevertheless, we will find that this commonality is only partial: A person and their humem counterpart complement each other, with each aspiring toward and achieving endeavors that are beyond the reach of the other.

Coming to Be—Prevalent Proto-Humems

Early kinds of EP consolidations already exist. We will see that, in a number of ways, these EP bodies already exhibit the beginnings of dynamic behavior. We can call these entities *proto-humems* to signify that, while they demonstrate a number of humem-like attributes, they still do not exemplify humems in the more complete sense. For instance, they are only partial consolidations of certain aspects of one's EP. As we will see, there is no sharp demarcation between the proto-humem stage and the realization of the fuller humem status; it is largely a matter of degree. As we progress in our discussion, however, we'll obtain a better sense of what characterizes, or makes up, freestanding and comprehensive humems.

---

[17] The term *agency* is sometimes used to describe the ability of an individual to act autonomously in both decision and action. We will discover the emergence of degrees of agency in humems, while still understanding that the constitution of the individual never implies any kind of absolute independence from the environment. Similar limitations, however, are valid also for agency as it pertains to people.

Mind Dwellers—Ancient Proto-Humems

As discussed earlier, historically, most people's EPs were bio-EPs—those EPs residing in the minds of others. For someone we know well, the composite of their EP in our mind does not resemble a static memory, like a picture or document; instead, it behaves more like an active emulation of a living individual.

Because this entity is an aggregation of their EP and can be shown to be dynamic in a number of ways, we can regard this emulation, or model, of a person as their proto-humem living in our minds. As we imagine or think about the person, we are in effect actuating or running this emulation. This is how we can recognize them. This is how we can see, hear, and feel them so vividly in our dreams.

Let's examine these models within our minds more closely. Notably, unlike a static memory, a model emulating someone's dynamics can manifest itself in circumstances that the originating person never actually experienced. For example, we can imagine the proto-humem speaking words that the original person never said. We can envision how the person would react were they to find themselves in a new situation. Most of us can do this quite easily when we think about close acquaintances. If you frequently have realistic dreams, and remember them, you can readily attest to the veracity of this phenomenon.

A clear demonstration of this ability to emulate and forecast the behavior of another person is impersonation. Inhibitions aside, most of us are able to mimic others with varying degrees of accuracy. A good mimic can convincingly imitate how another person would behave in an imaginary situation—from their voice, facial expression, and body language, to their choice of words, and even their probable answer to a question.

This is not merely replaying a splicing of previous recordings but rather applying one's existing knowledge of another individual to forecast how they would sound, look, or express themselves in a new situation. Such mimicry is an externally apparent manifestation of this internal model of the other person—their proto-humem—living in one's mind.

The mechanisms of recognition also illustrate some of the workings of these proto-humems. Is recognition a retrospective process—that is, are we simply comparing new stimuli to preexisting static data in our brains? Do we first listen, see, and smell, and then compare this to what we have in our brains while searching for the best match? Well, maybe that's sometimes the case, but in many instances it seems that a more proactive and predictive mechanism is in progress. When I search for someone—say, my child—in a crowded park, it is clear that a more sophisticated process is at play. I am not looking at, and fully "absorbing," every person in view and then matching this input with the contents of my memory. Rather, I am invoking a model of my child and projecting how she would look in the given environment. The model in my mind allows me to customize the "what" that I am looking for, and only then do I start actually searching for a match.

This emulation has relatively static components, like her physical features and the color of her hair and clothes, but it also has many dynamic factors, like her gait, her gestures, and how her hair moves in the wind. Then, when observing a group of children running on a grassy slope in the distance, where my child has never run before, I create a moving picture in my mind of how she would look if she were running there now. Thus, I'm able to quickly find her in the midst of a large group of other children because I already have a good idea of what I am looking for in that particular setting. Similarly, before finding her, I can quickly dismiss, or filter out, the images of other children without having to first inspect each one in detail.

These proto-humems that "occupy" our minds have many other lifelike characteristics. For example, they grow and develop if nurtured. Ongoing interaction with the corresponding (originating) person is the most common way of developing the proto-humem; that is, simply getting to know the person better. If the proto-humem is neglected, then it slowly declines; we commonly call this "forgetting the person." Without the help of modern media such as photographs and video, even most parents are unable to invoke clear images of the way their children were many years previously. This is presumably because the child's proto-humem in the parent's mind grows and develops in tandem with the child. By contrast, a parent who has experienced the death of a child may, for many years hence, be able to remember them as they were in life, before dying, because the child's humem, frozen in time, remains forever young in the parent's mind.

We often have what we call "imaginary" conversations in our heads with our acquaintances. Although we mostly realize that these models of others in our minds are wholly contained within us and are presumably part of us, we still often perceive them as separate and independent entities. Using our new terminology, we can resolve this cognitive oddity by saying instead that we are having "actual" conversations with their proto-humems in our minds. Similarly, it is not unusual for someone to say to us, "When you appeared in my dream last night, I did this and you did that, I said this and you said that." But what do they mean by the "I" and the "you"? Of course, we know that this is all happening inside their heads—there is no one else. Yet, at the same time, we often empathize with our proto-humem in the person's dream and may even feel responsible for its behavior.

Perhaps the most poignant example of having a proto-humem living in one's mind is the experience of being in love. For those who have had the delightful luck or the brutal misfortune of this singular experience, the veracity and power of the phenomenon is absolutely clear. Here too, it is obvious that the alluring proto-humem in one's mind can exist quite independently from the originating person, who may not even be aware of its presence.

Yet another lifelike attribute of these proto-humems is their ability to replicate and grow, not only by direct seeding and nurturing by the originating person, but also via intermediaries such as other people and abiotic media. A key point here is that these representations of others in our minds are not like static archives of memory. These proto-humems behave like living and dynamic beings—over long periods, they either decline, or they grow and adapt to the times. As we saw earlier, abiotic-EP is becoming the dominant part of our total EP, with the bulk currently being created through digital media. However, this massive expansion of the abiotic-EP is also opening new growth opportunities for the bio-EP; it allows us to increase our global reach into the minds of others.

Instead of being formed via direct interactions with other people, the proto-humems in our heads are increasingly originating from abiotic-EP through modern communication channels.

While the examples described above should be familiar to most readers, this view of the primordial bio-EP-based proto-humems that dwell in our minds provides a useful benchmark for studying the characteristics of the abiotic-EP-based humems—those humems that exist outside our brains. We will focus on the abiotic kind of humems for the remainder of this book.

A Book's First Pulse—Traditional EP Transforming into Modern Proto-Humems

When we read a book—say, an autobiography—portraying a deceased person's character, we may experience it as an intimate interaction. Receiving a personal letter from a loved one, posthumously, may invoke similar feelings. We may be profoundly touched, moved to fascination or tears, and we may feel as if the writer is physically present and addressing us directly.

What is actually happening in these situations? *Someone* is communicating something. Clearly, it is not a person who is interacting with us in any bodily way. Employing our new terminology, however, we can consistently argue that the book, as a consolidation of part of a personal EP, is the author's proto-humem. It follows that it is this proto-humem that communicates with us as we read, and subsequently becomes alive in our minds.

To get a better feel for why we can call certain books proto-humems, or how books may be seen to possess a number of humem-like characteristics, let's consider more closely what we mean by a book. When we are asked to imagine one, we may naively think of a rectangular object composed of bound paper pages. Yet, in contemplating a specific book—say, Homer's *The Odyssey*—the object of paper and ink is clearly not *the* book but rather a *copy*, or *manifestation*, of the book. *The Odyssey* originated thousands of years ago, most likely as an oral tradition that was later transcribed into written and other forms. It is not a fixed material object, but a continually developing and adapting body of information, or story, that has been and continues to be conveyed in multiple interrelated forms such as sounds, print, images, and movies. It resides on a wide variety of media such as brains, paint on clay vases, stone sculptures, ink on paper, and digital storage. While a specific copy of *The Odyssey* may appear fixed and static at a given moment, *The Odyssey*—as a story—has been highly dynamic over the period of its existence, or lifetime. A paper copy of *The Odyssey* in relation to its full story is analogous to a snapshot of a person in relation to their entire life.

Like any widely published book, *The Odyssey* is so dispersed that it makes little practical sense to ask where it is located. Yet, it is completely real: from a material perspective, it does ultimately reside on a finite number of carriers, such as specific people, vases, books, digital copies, and so on. And each of these has a well-defined location at any given time.

Moreover, we see that despite its apparent nebulosity in material composition and location, *The Odyssey* has a distinctive character. The central parts of its story are widely recognized as being associated with its special identity. When presented with a snippet of the story, a well-read person can readily identify its source. Also, in present times, it makes little difference whether *The Odyssey's* hero, Odysseus, was a real person or not. Those who know the story recognize his character and may identify with his tribulations.

These attributes of being physically dispersed but conceptually unified, and being dynamic and developing but still retaining a core character and recognizable identity, exemplify many of the fundamental humem characteristics that we discussed previously. Certain books have exhibited such qualities for extended periods. With emerging technologies, these traditional forms of EP are developing even more vibrant and lifelike behaviors. They can more rapidly adapt to present circumstances and begin to merge with other facets of their authors' characters. Let's examine how this is happening and how it may unfold.

Let's proceed by replacing the image of a printed book in our minds with the modern alternative of a digital version together with all the attendant capabilities of the devices that render such books to readers.

In such a view, the book proto-humem's memory contains not only the original book but also much more. In addition to the enhanced functions provided by the computing device, any information relating to the book that may be accessed on the Web (reader reviews, plot synopses, author video interviews, to name a few) can also be considered part of this proto-humem's memory. In fact, each word and phrase that it contains is now dynamic and interconnected to a wealth of additional information. The story as a whole has new abilities and modes of expression.

Recently, these book proto-humems have started speaking, translating, and listening. Over time, text-to-speech recitals have improved and become quite useable. They can be asked to speed up or slow down, or to pause or resume the narrative. Likewise, while real-time language translations are rudimentary at the present time, they too will undoubtedly progress. By using voice recognition technology and that of emerging mobile digital assistants, for example, one's interactions with the proto-humem can become more similar to inter-human communications. When interacting with *The* Odyssey proto-humem, to return to our example, one may ask a question about the geography of Ithaca or how to pronounce Penelope's name in Greek. The answers may be presented in the form of text, sound, images, video, or a mix of media most appropriate to the situation.

Here something fundamental is changing. The traditional act of reading a story is now transforming into a conversation with the story. The reader is no longer just a reader but also a communicator and participant. (Still, I'll use the term "reader" for the remainder of this discussion.) Already, the information in these actions is beginning to flow in both directions. When one buys an e-book, for example, one's perception of the story may be influenced by reader reviews. These have become part of the bigger story. That is, not only do the readers absorb the story, but the story also absorbs their reactions.

Perhaps the most compelling feature of a book proto-humem is its integration with the rest of the author's humem (the totality of the originator's EP). Such a merger facilitates untold enrichment of the parts by the whole. For example, if the "main" humem includes the originator's voice recordings, which are not necessarily related to the book, then voice-recognition technology can learn how the author spoke and thereby emulate their speech. This can be applied to allow the book to "express itself" in the author's voice even if the author never read the book aloud. Furthermore, when the book humem is asked a question, it can also refer to other parts of its own knowledge beyond the information contained in the book. These include the author's other books, interviews, videos, emails, and so on.

Often, the humem may not know the full answer. Then it may consult the global information base of all the world's publicly accessible knowledge. (Presently, the World Wide Web via the Internet is the predominant such medium.) Here we need to make an important distinction. In this vision, where the humem is more mature and advanced, it does not simply redirect the reader to an external information source. Rather, it does something much more powerful. It retrieves external knowledge and processes it in the context of the narrative, and customizes the response for the reader. Furthermore, it distills this information in a way that is consistent with its character—in its own voice. This is just what a person would do in such a situation.

Taking this a bit further, let's imagine for a moment that the originator, or author, is holding their book and sitting next to the reader. Also, let's imagine that the originator has unlimited time and patience, like a book or machine, and also knows a lot about the reader, as a person may know a friend, a family member, or a child. Imagine also that the originator is up to date on current world affairs, especially on topics relating to the narrative of the original book. How would the story and the answers to the reader's questions sound now?

Lets consider how the originator's humem could facilitate such a scenario. Assuming the reader also had an associated humem that was able to interact with that of the originator, a mutual knowledge could be established. Thus, in the way that a child asks a parent a question and the parent supplements their own knowledge with reference material and adapts the answer to the context and the child's intellectual level, so can the originator's humem, after getting to know the reader's humem, adapt the answer to the reader. Of course, an answer to a question is only one of many possible exchanges; this paradigm applies to other types of dialog as well. It depicts interactions between dynamic humems and people—each responding to the essence of the other, and, not less importantly, to the backdrop of the current world in which they coexist.

These examples describe entirely feasible scenarios; nothing here is beyond the current technological horizon. While existing implementations may still be basic, they are steadily progressing and becoming more useable. Thus, the emergence of such systems is inevitable. It is mainly a question of how soon they will attain the sophistication to be widely adoptable. As consolidations of personal EPs mature, the resultant humems will become ever more predominant; and then modern types of books will be just another of the humems' multiple modes of interaction.

We should keep in mind that these systems could be designed as supersets of existing forms of information and communication. The legacy components can still persist as part of the consolidated bodies. Thus, traditional expressions can still be possible. In relation to the example above, a book can still manifest as a paper book for those who desire the traditional experience—a black-and-white version for when the colors become too bright.

This depiction of a book exemplifies the key point relating to the transformation, or metamorphosis, of older but well-maintained EP in the future. As previously inert parts of the EP are combined and immersed in the fertile protoplasm of emerging technologies, the whole starts becoming alive in remarkable ways. Then one may hear rustlings on dusty bookshelves and feel an old book's first soft pulse in one's hands.

Our Storefront Reflections—Accidental Proto-Humems

Proto-humems are emerging in unexpected places and in astonishing forms. As an example, let's look at an individual's online shopping account. We may find that it embodies much more than what first meets the eye. Perhaps like you, I enjoy the convenience of online shopping. Furthermore, wherever possible, I like to make most of my purchases at a single store (or just a few stores). This preference is based on apparent advantages, such as a narrower distribution of my payment and personal information, my familiarity with the store's shopping process, customer loyalty incentives, good customer service, and a flexible return policy. At the time of this writing, I've had an account at my favorite online store for more than a decade.

Since this retailer partners with many other sellers, it offers a broad range of products at competitive prices. Looking at my online account, I see that my purchases include books, electronics, clothes, sporting goods, tools, medications, food, and many other kinds of merchandise. When I consider how much each of these categories reveals about my life, it becomes immediately apparent that this online retailer can know a great deal about me—probably a lot more than I would want any stranger to know.

For brevity's sake, in this discussion let's call the automated agent that interacts with me and knows my profile and purchasing behavior "the machine."

I search for products within the store and read ratings and reviews written by other shoppers. Thus, I create a detailed shopping-browsing history. The machine knows not only what I bought but also what I considered buying. It has observed me deliberating between options and has seen which ones I chose. Consequently, the machine knows a lot about how I make decisions. The machine does not always have to think very hard since I can explicitly record my product interests using a feature called a "wish list." There are many reasons why I may save an item in the wish list and then either buy it later or discard it. Moreover, I can write reviews and post pictures and videos in the online store relating to my experience with the products. These actions all provide additional windows into my interests and decision-making behavior.

The machine knows my shipping and billing information and how they have changed over the years. It knows the addresses and names of people to whom I have sent gifts. This, also, provides a wealth of personal and social insight. However, the machine can go much deeper into my personality: the nature of specific products considered or purchased provides intimate information about me. For example, if I consistently purchase a particular category of literature, it is likely that I am reading these books myself. And if the machine has access to the contents of the books, which it often does, then it has an idea of some of the specific contents of my memory.[18] It can deduce, for example, that I am most likely acquainted with a certain story or subject. Looking at the list of my recently purchased books, I see that the machine can infer details relating to the content of my mind that even some of my closest friends could not know. (They have not read those particular books, and do not necessarily know that I have read them.)

As another example, by observing the clothes I buy, if these are consistently sent to my home address and match my gender and size, the machine can conclude, with high probability, that these are for my own use. The machine can then also know something about my body dimensions and how these change over the years. Correlating this information with other purchases such as sports equipment, medicine, and food, it can understand trends in my health and lifestyle that even I may not be aware of.

The machine, from the knowledge gleaned through my preferences, behavior, and purchases, often proposes things that may interest me. Subsequently, my response to these proposals provides feedback to the machine on how well it is beginning to comprehend me. If I am seduced by its offers, it knows that it is progressing in understanding my tastes and tendencies. Conversely, if I ignore its proposals, it can conclude that it should consider other approaches.

Whole books could be dedicated to the implications of this situation. We, however, have covered more than enough to illustrate the intended point: that the online shopping account meets our criteria for a proto-humem. It is a consolidation of numerous aspects of a person's character, including their likes and dislikes, ways of thinking, modes of

---

[18] I am not asserting that any particular online bookstore currently looks at the internal content of books and uses this to profile a customer. I am saying however, that these commercial entities already do conceptually similar things and are quite capable of refining such analyses if this is found to further their purposes; the raw data is there.

decision-making, and life history.

Moreover, this proto-humem displays multiple levels of dynamic behavior. Due to advances in the underlying technologies, it is constantly improving and expanding its abilities. It "thinks" together with its human counterpart, makes suggestions based on prior behavior, and is influenced by the broader environment. For example, if it knows, thanks to a rating or review, that one enjoyed a book written by a particular author, the proto-humem can take note and then, when this author writes a new book, it can recommend it. Furthermore, if another author, unknown to the person, writes a book that has been deemed by others to be closely related to the previous author's work, then, here too, the humem can bring this information to one's attention. We see that the proto-humem could be quite competent, for example, in making a choice of gift for the person. It could even be quite obnoxious: by noticing an upward trend in the person's clothing sizes, it could recommend a diet book or a low-fat variety of a previously purchased type of food. Thus, we can plainly see that this proto-humem can exhibit behavior of the type that could emanate from a friend, an assistant, or a caring spouse.

Many feel somewhat uncomfortable with this situation. A primary source of this unease is the realization that although this proto-humem may, in some respects, be able to emulate a friend, it certainly is not one. Even if we could forgive it for being a machine, a much more fundamental problem remains: this humem does not have our best interests at heart. Simply stated, it is owned by and subservient to a commercial entity whose main mission is to sell us things for profit. We can call this proto-humem an *accidental* humem[19] in the sense that when we first created online shopping accounts, almost no one, neither the customers nor the retailers, had much idea where this would lead. Even today, very few consumers who sign on to such services have an in-depth understanding of the full implications of these engagements.

_____

[19] *Incidental* or *unanticipated* would also describe this kind of humem quite well.

The crux of the issue is that the online store system is very useful, and at present no good alternatives exist. Thus, with current methods, convenience goes hand in hand with the loss of privacy, control, and ownership of one's data. Later we will more closely examine these and other forms of abiotic proto-humems, and consider what should be done to ensure that our personal humems are created within the right frameworks, under the appropriate control, and with the correct alignment of interests with their human originators. We will see that properly established humems can provide even greater convenience than these commercial, accidental ones, while retaining much higher levels of individual privacy and control.

Connected We Stand—Social Proto-Humems

One's email account, personal webpage, blog, and video-sharing channel are all aggregations of one's abiotic-EP. Currently, however, one's social network account best exemplifies a proto-humem by exhibiting the beginnings of dynamic and semi-independent behavior.

We can call each such account, corresponding to an individual person, a *social network instance.* When actively maintained and cultivated, an instance may emulate an individual's "real-world" presence by forming a comprehensive aggregation of their EP. Notably, the social network instance does not stand alone: it is linked to similar entities that represent other people, thereby mirroring one's "real-life" connections.

Popular applications on social media sites display increasingly sophisticated dynamics in which some actions are performed autonomously. While an instance mimics a person's life and relationships, it also has an inherent vitality and generates relationships that could not have existed before the advent of social media. It reflects a person's activities in the physical world, but possibly more significantly, it also influences a person's future actions. The integrity and social standings of the social network proto-humems are directly shaping those of their physical human counterparts. They affect and often direct what we do, and also when and with whom we do it.

These social network instances are starting to behave like living things—in some ways, arguably, similar to pets or children. They need—and even ask—to be nourished to maintain their health and luster. If neglected, they wither and fade. Following, and increasingly during, social events, we feel obliged to feed these social networks with commentary and photographs. Habits form: the more we nurture them, the hungrier they become. The symbiosis between these proto-humems and us is becoming stronger. Already, for many of us, if a physical "real-life" event does not have a social network counterpart, there is almost a feeling that the event never happened or was not worth the effort. [20]

Out of all the abiotic proto-humem examples presented so far, the social network is probably the most relationship-centric. Typically, when one creates a new instance, or account, the application starts by mirroring their "real-life" connections and relationships—their social map. If one is new to the social network, but has family, friends, and colleagues who are previously established in it, a significant part of the new member's EP already exists in the network. Many of one's memories and experiences are present in the social network instances of their contacts, and are inherited as one defines their relationships to others. Thereafter, changes or growth in each social network instance permeate to those around it.

We will examine relationships in more detail in chapter 4. While the mirroring of real-life relationships is a starting point, we will find that new kinds of relationships are also emerging—relationships that will play a central role in the future dynamics between humems and people.

Social media applications are becoming an increasingly important part of our lives. Since their inception, however, these developments have given rise to concerns relating to ownership, control, and privacy. We will discuss these issues further in chapter 6 when we explore the notion of humem rights and citizenship.

---

[20] This type of behavior is reminiscent of the thought experiment described by Daniel Kahneman and others in which people greatly devalue a pleasurable experience if they believe that they will have no recollection of the event at a later time.

Benefits of EP Consolidation

Why is a combined EP better than a dispersed EP? How does consolidation affect the EP's prospects for growth and longevity, and its ability to influence the world? To answer these questions, let's look at some immediate consequences of the formation of humems through the amalgamation of various components of individuals' EPs.

Better Together—Complexity Increases Capability

Numerous experiments, computer simulations, and scientific theories demonstrate how an agglomeration of simple parts, each apparently well understood as a discrete entity, often leads to outcomes that were not or could not have been predicted previously. Such a merging can result in a level or type of complexity that renders the whole greater than the sum of the parts and bestows the resulting entity with properties that the parts alone did not possess to any discernable degree. The discipline that describes such systems is commonly called the Emergence Theory of Complex Systems.

Examples of this phenomenon have been discovered at many levels in numerous domains including physical non-living systems, life processes in the natural world, thought processes, and the dynamics of societies. In the present context, I argue that the humem as a whole can possess numerous characteristics that could not exist to any significant degree if the fragments of the EP remained as such. This effect is apparent in many of the humem examples that we observe.

Mutual Insurance—Consolidation Enables the Survival of Parts

A valuable and sustainable whole enables the survival of some of its parts that could not survive alone. Here, in particular, I am considering a whole that contains components that could conceivably exist separately for a time but would not endure alone for long periods. Such atrophy could be due to a susceptibility of the parts to become temporarily irrelevant or worthless in certain settings or at certain times.

In composite systems, some of the parts may not have a significant current value or a currently appreciated future value. However, these same components may eventually turn out to have immense worth at some time in the future. Combining this future value factor of the parts together with the one relating to the emergent properties of complex systems, we see that maintaining the aggregate over time can facilitate the emergence of valuable future entities, now unknown.

We see examples of similar workings in widely disparate systems such as the bodies of organisms, libraries, and many others. It appears that at many levels within organisms some components exist that are not apparently useful or functional at the present time. For instance, a substantial amount of genetic material may be inert remnants, without current functions or expressions. Recently it has been found that much of what was previously designated as "junk" or "non-coding" DNA does in fact have important functions. Still, it appears that certain parts truly are unused or unexpressed. Some constituents may be "true junk" in the sense that they will never be needed. Yet, like a desert-dwelling people's inherent ability to learn to swim, other dormant components may eventually become useful in future environments—a kind of biological insurance policy.

A comparable situation prevails in libraries and archives where, for a finite time into the future, the bulk of stored material may not be useful but a smaller portion may turn out to be extremely valuable. Because of our incomplete understanding of current systems and how they will change in the future, it's impossible to differentiate between the more and less important parts. Nevertheless, the potential usefulness of even a small percentage commonly offsets the cost of retaining the whole. Keep them all and they'll provide great returns as a collective, even though many individual books may not contribute at all. From the books' perspective, this is a form of mutual insurance.

Moreover, some types of EPs may even have a *negative value* at certain points in time. By this I mean that the entity assigning the value would prefer that the EP not exist at all. Some kinds of political, histori-cal, and religious material may be regarded as offensive in certain settings. If they existed as loose units, they might be destroyed.

But if those negative parts are inseparable or difficult to detach from those that are deemed valuable, they can "tag along" into the future, when they may become precious. For example, we can imagine certain segments of a Greek philosopher's manuscript being considered subversive and worthy of deletion at certain points in history. Yet, solely due to the fact that the book was bound and glued together, the positive value of the other sections may have ensured the survival of the less desirable parts.

Parts of the fragmented personal EP are often prone to deletion or loss due to their unknown or unappreciated future value. By contrast, consolidating the EP pieces into a robust whole humem imparts resiliency and longevity to all its parts and facilitates the emergence of future value from the constantly growing collective.

Essentially Together—Humem Body Organs

We have seen that an aggregation of units can improve their prospects for survival and can lead to the emergence of novel characteristics from the resulting whole. These units could, at least for a time, also exist separately. However, associations of units can give rise to new parts and subsystems that could not exist in isolation at all.

This principle seems to hold true for most systems that surpass certain levels of complexity, as well as for many forms of symbiosis in which apparently separate systems have a critical interdependence. Numerous examples exist in the bodies of living organisms at all levels, from the organelles of cells to larger structures such as organs in higher organisms. These components have functions that are only relevant in the context of the system. Moreover, in various other domains, aggregations result in the emergence of entities that are often completely absent in the separate units and meaningless without the whole. For instance, societies give rise to unique organizations and hierarchies, features that are not necessarily present to a lesser degree in the individual people of which they are composed.

Similarly, with the convergence of the EP into complex humems, we can reasonably expect the growth of subsystems and abilities that would be inconceivable and irrelevant within the dispersed EP. Over time, some of these formations may become crucial for the proper functioning of the humem as a whole. In this sense, these dependent and necessary subsystems contained within humems can be thought of as humem body organs.

Furthermore, aggregations of humems will lead to higher-level configurations, mirroring and sometimes diverging from those of their human counterparts within societies.

One of a Kind—Complexity with Variance Increases Individuality

The formation of the humem increases the uniqueness, or individuality, of the personal EP.[21] As we will see, this is an outcome of an increasing complexity of the whole in conjunction with a greater degree of variance of its parts.

Complexity has varied connotations in different disciplines. In the current context, we can view the complexity of an entity in the simplest terms possible: the more parts, and the more modes of interaction among the parts, the more complex the entity. For example, a mechanical watch with multiple parts and more interactions among them is more complex than a watch with fewer interacting parts.

However, regardless of how intricate and complex a watch may be, it is not necessarily unique. Typically many identical watches exist, each with the same parts and interactions among them. Thus complexity in itself does not ensure uniqueness, or, in our context, what we may call *individuality*.

Now let's imagine that the watch has assembly options. For instance, the watchstrap can be made of either metal or leather in various colors. Additionally, the watch face can be of different textures and backgrounds. Perhaps some internal parts are also customizable. We can call the number of different options of a part its *degree of variance*. As the

---

[21] I assume here that individuality or uniqueness is a desirable EP trait as it is for people. However, some people may object, and claim that it is not axiomatic to call individuality a positive trait in any absolute sense. Here I do not argue this point.

number of customizable parts increases, and as the variance of each of these parts increases, the number of possible unique watches increases too.[22]

So, although complexity alone does not imply uniqueness, complexity in the sense of a multitude of parts coupled with a high degree of variance of the parts does lead to an increased potential for the uniqueness of individuals in a species. (In the example above, we can regard the series of watches with interchangeable parts as a species of watch—their parts can be mixed and matched and they will still work.)

For people, this notion of individuality is expressed at multiple levels, essentially including every aspect of our being. Our individuality is affected by our measurable attributes—microscopic and macroscopic—in the material domain, and, by currently less quantifiable traits, such as our behavior and the contents of our minds.

Physically, we are all composed of the same substances and are chemically and genetically similar to each other. When examined at a sufficiently small scale, no part of us is unique. For example, virtually all of our genes have copies contained in the bodies of many other people. In particular, genetically, we are a combination of the specific genes of our parents. Thus, none of the fragments of our genetic makeup is special. However, due to the vast number of genes and the variation in the way that they are joined and expressed to make an individual, the specific combination of which we are composed is almost certainly unique.[23]

As a macroscopic extension of this idea, consider our external appearance. Somewhere out there is someone else with an almost identical nose to ours, the same shape of chin, very similar eyes, and so on. But the chance of all of these similar parts—this particular combination—existing in another person, in a way that could confuse our close friends, is practically negligible. In this sense, we are each quite unique. Viewed broadly enough, when sufficient numbers of parameters for

---

[22] For simplicity we can assume that the "choice" of an option for one part is independent of that of other parts.

[23] Except perhaps for identical twins but even they differ at other levels.

comparison are included, we are each absolutely unique in appearance.

Analogously, most of our experiences and memories are composed of parts, or fragments, each of which is not particularly exceptional. We have probably never heard a word or sound or song, nor tasted a food, nor seen a sight, which has not already been heard or tasted or seen by someone else. We have most likely never experienced a pain, nor felt a delight, which has not already been experienced or felt by someone, somewhere, sometime. However, our particular mix and sequence of experience is exclusive to our lives.

It is the singular blend of these various constituents of our being that forms our individuality.

In the modern world, there is a much greater variety of things we can do and think about than there was in the past. Thus, we have an ever-increasing potential for individuality in action and thought. In earlier times, the character of people in a typical community was once much more homogeneous. Take, for example, two similarly aged persons living in a small village in medieval Europe. Likely illiterate, never having travelled more than a few miles from their village, they would have been much less individually distinct than two modern city dwellers. They probably would have seen the same landscapes, known the same people, heard the same jokes, played the same games, and eaten the same kinds of food. Their minds would have contained similar memories and ideas. In other words, their material and conceptual worlds would have been much more uniform than those of modern people. They would have been unable to conceive as much independent thought because they would not have possessed enough diversity of experience.

The variance in modern populaces is clearly much greater. Especially in big cities, physically adjacent individuals can be very different in many ways. The bus driver and the physics professor who meet most mornings, before parting ways at the university bus stop, may differ immensely. Their activities, their thoughts, their dreams, and the nature of their interactions with others likely separate them at many levels.

Let's now extend this notion to humems. The more a humem grows, the more components or units of information it contains, and the greater the variance of these individual components, the more unique it becomes. The consolidation of the otherwise indistinct EP fragments into a singularly interconnected whole produces the unique humem-individual.

Human bodies and brains have not structurally changed much in recorded history. Thus, there has not been a significant advance in the potential for people's bodily variance. However, the potential for variance in the content of our brains, or what we may call the *character of our minds,* has grown dramatically in recent times. These developments are mirrored and amplified in our rapidly expanding EPs. Moreover, our interaction with the world is increasingly becoming an EP-centric process. Thus, the coalescence of the EP into humems implies that these entities will become substantial carriers of our characters and expansions of our individuality. And the richer and more sophisticated the humems become, the more our scope of expression as unique individuals multiplies.

Unification of the Personal EP

Presently, personal EPs are typically composed of widely dispersed fragments intermingled with clumps or localized consolidations of EP, the most substantial of which we can call proto-humems.

The proto-humem instances described earlier are but a small sampling of the rapidly growing variety of these emerging entities. Other common examples of one's abiotic-EP clusters include the data contained in one's email account, the information contained in one's bank account or credit card records, one's data on a personal blog or video sharing website, one's health records held by a medical institution or insurance company, the records of one's call history and text messages held by a mobile-phone service provider, and the personal data within one's computing devices.

Clearly, people have consciously designed much of the technical infrastructure that enables proto-humems to exist. But if we observe the bigger picture and the often-unforeseen convergences of various paths of technological progress, it seems that many types of humems are emerging spontaneously. In fact, proto-humems, as they exist today, are almost always quite different from what the designers envisioned when they created the first prototypes.

Convergences in EP-related abilities and behaviors are happening at multiple levels. For example, although the social network and online store proto-humems, presented earlier, started out with quite different functionalities and behaviors, over time their overlap has been growing. The online stores are increasingly social, and the social networks are increasingly commercial. Another well-known example of this convergence, relating to EP processing mechanisms, is that of the mobile phone and personal computer. Although they had distinct functionalities in their early days, they are rapidly merging into an extended family of devices with similar behaviors and uses.

Each of the personal EP collections reflects complementary facets of a person's memory, experience, character, and behavior. The separate EP parts are related in a host of intricate ways—a wealth of information and relevance resides in their interconnections. However, due to the current compartmentalization of these data sets and their related applications, many of these links are broken and much of their richness lost. For instance, at this time, the proto-humems of social networks are mostly disconnected from the knowledge bases of genealogy and legacy applications, even though the former constitutes the roots and substance of the latter. Personal and professional abiotic-EPs are also formally separated, while in practice the activities relating to these parts of one's life overlap significantly.

Moreover, our minds do not separate these various facets of our lives. In our brains, the counterparts of these diverse types of abiotic-EPs and proto-humems are stored together, inextricably intertwined and mutually enriching. In the abiotic-EP realm, much is forfeited as a result of these unnatural divides.

The establishment of a comprehensive personal humem by gathering together this diaspora of abiotic-EP and proto-humems is vital for cultivating these interconnections. This consolidation provides the humem with the cornucopia of ingredients that are essential for the formation of a whole individual.

The Birth of the Personal Humem

In this chapter we have viewed humems from a number of complementary perspectives. First, we regarded a humem as a coalescence of personal EP into a body with character and identity. We noted, however, that the body's boundaries are not sharply demarcated. I illustrated this notion by modeling the humem as a concentric sphere with a clearly affiliated center and gradually more nebulous peripheries. Then, I suggested that the ambiguity in the definition of the humem is similar to that of the person, or personhood, in its broader, modern sense.

The concept of personhood is constantly changing and developing, and is dependent on multiple factors, including jurisdiction, period, and circumstance. Nevertheless, its meaning is mostly intuitive and the term is also useful and necessary for a wide range of practical applications. I surmised that, in the future, the concept of a humem will become similarly recognizable and indispensible.

After examining a few existing examples of personal EP clusters, or proto-humems, we considered some of the conceptual advantages of EP consolidation. Finally, I reiterated that the whole personal humem can be understood as the compendium of one's abiotic-EP—especially one's more clearly affiliated EP, including those nascent consolidations called proto-humems—and the relevance contained within the interconnections among the parts.

Yet, for many, the picture may still seem somewhat obscure. For instance, how can we envision a whole and consolidated personal humem arising from the motley assortment of proto-humems we now see all around us? One way to think about it is to view the proto-humems that are affiliated with a specific person, not only as components of a single encompassing personal humem, but also as its various manifestations. Put differently, instead of regarding these proto-humems as different entities, we can see them as diverse expressions of a single underlying entity. We will further develop this concept of a physically dispersed but organizationally consolidated humem starting in chapter 5.

At this point, one may also ask, "How do we actually create, or establish, a humem?" The answer to this question too will become more intuitive as we progress. For now it may be helpful to compare it to how a state citizen comes into being. What does it mean to create, or establish, a new citizen?

First, let's take one possible source of confusion off the table. Although the terms are often used interchangeably, citizens and people are not the same thing. That is, in modern democracies at least, citizenship is not equivalent to personhood. (In certain oppressive regimes, where populaces are sometimes treated as state property, the distinction may be somewhat less clear.) Citizenship is more like a membership, which a person may or may not hold. Thus, the birth of a baby is not equivalent to the creation of a citizen, even though in practice these events often coincide. Babies can be born without being bestowed citizenship, and persons holding a citizenship can relinquish or lose this affiliation later in life. Nonetheless, in nation-states, the birth or immigration of a physical person is a prerequisite for the "making" of a new citizen, and the person's death subsequently results in its termination.

From a formal standpoint, the making of a new citizen is almost trivial in simplicity. In essence, at the time of its inception, almost nothing happens beyond the creation of a new entry in a ledger or database and perhaps the generation of a few certificates. These, of course, do not constitute the event of creation itself but only a record of the event. In one sense, the actual event can be viewed as an almost completely virtual occurrence since nothing materially changes. Yet, in another sense, at the moment the citizenship is established, the associated person undergoes a profound change. They adopt, or inherit, a vast framework and long history of existing rights, obligations, cultural conventions, legal precedents, and so on. Moreover, they assume the ongoing implications and developments of the citizenship—typically for the full duration of their life.

Although the establishment of a humem will probably be less contingent on the existence of a specific physical entity, or body, in most other respects it will be similar to that of a citizen. For instance, depending on the humem jurisdiction or habitat (the environment in which the humem will be created), the candidate may be subject to a number of criteria, such as having a relationship with existing humems, having supporting sponsors, or perhaps even having a minimum monetary worth or commitment to investment in the hosting environment. (Similar criteria are also de facto paths to citizenship in many prominent nation-states today.)

As we'll see in later chapters, once an appropriate environment exists for humems, the creation of a humem will be conceptually similar to that of a national citizen. At its inception, the new humem will inherit a broad framework of preexisting attributes, rights, protections, obligations, and codes of conduct—all conferred by the humem jurisdiction in which it is established. These, combined with the humem's individual features, will constitute its basic nature.[24] As with differences between nations, there will be variances among different humem jurisdictions. Thus, the character of humems and the modes of their establishment will develop in tandem with the progress of their hosting environments.

---

[24] Later we'll see that humems will also possess the equivalent of national citizenship—humem-citizenship. So as with personhood, humemhood is not equivalent to humem-citizenship. Since people (and human rights) are universally recognized, it is possible to separate between personhood (or world citizenship) and national citizenship. But, as with newly emancipated slaves, in the incipient stages, the forms of humems' citizenship will largely define the character of their humemhood. Once humems also achieve a more universal recognition, it may be possible to more clearly differentiate between their humemhood and their formal standing in specific humem jurisdictions.

CHAPTER 4

RELATIONSHIPS AND ALPHA-PAIRS

This chapter examines the relationships between and among people and humems, and explores how these interactions will occur and be regulated. As we will see, while humems emulate the types of relationships that occur between people, humems' distinctive characteristics may also give rise to novel associations that do not have counterparts in traditional interpersonal relationships.

In using the word "relationships" in regard to humems, I mean all that this term signifies for people and society—not a metaphor, but rather the most appropriate and natural description of these associations.

Humem Interactions

Before examining the nature of humems' relationships, let's briefly discuss the mechanisms of the interactions themselves. When we think of people and how they interact with each other, with society, and with institutions, we mostly have an intuitive sense of how this happens. This intuition stems from our vast personal experience with relationships and our constant immersion in social environments. However, as we prepare to study humem relationships, we may feel unsure of how humems will behave and interact. This may lead us to think that we cannot properly study the fundamentals of humem relationships without a more complete knowledge of the means of their expression. But if we consider how the methods of human intercourse have trans-formed over time—even while underlying human relationships have remained essentially constant—we quickly see that the same can be true for humems.

For instance, the non-physical means of expressing romantic love have evolved over the millennia even though the basic impulses and implications of amorous relationships have stayed the same. In non-technological and illiterate societies, direct verbal and musical commu-nication was the mainstay for ages. Over time, this was expanded into letters of written words and poetry, which expressed the very same things and evoked the same primal emotions. More recently, instanta-neous electronic communications, including images, video, text, and ready-made emoticons—increasingly mediated by our social networking proto-humems—achieve similar results.

Likewise for humems, however rudimentary their initial communica-tion abilities may be at this time, lasting relationships can be estab-lished. Going forward, the mechanisms of their expression will certainly develop and change. No great leap of vision is required to gain a sense of how humems may interact in the future; as we saw earlier, we are already communicating with proto-humems and they are communi-cating among themselves. As they emerge, these new kinds of interac-tions quickly become second nature. And even though we can partially imagine how these interfaces will develop, the subsequent outcome will probably be much stranger than our boldest projections. Still, when these methods of interaction materialize, they will seem as ordinary to us then as sending a text message is today.

Here a brief thought experiment is useful: Let's imagine some time-transported people from the eighteenth century being shown a group of modern teenagers participating in a multiplayer, online battle game. They would see the kids in front of their screens simultaneously activat-ing game controllers to direct the actions of their virtual characters, communicating in voice to coordinate their moves and tactics on the battlefield, and interacting by text and graphic interfaces to allocate resources such as ammo and armor—all this while being physically separated by half a planet! In observing these interactions, the time-travelers would probably lose all confidence in their prior understanding of human behavior. While the kids' shrieks of delight would be familiar and timeless, everything else would appear more bizarre than in the voyagers' wildest dreams. Yet today's teenagers embrace these innovations so eagerly that not long after one appears, they seem unable to conceive of a world without it.

In the example of the digital game, we can consider the characters, each representing a human player, as a type of proto-humem. Their behaviors reflect the merging of their built-in, or encoded, functionalities with the directives and personalities of the human players. As gaming technologies progress, these types of proto-humems become increasingly human-like and personalized, and the component that more closely represents the human player's personality also grows.

In this kind of interplay between people and proto-humems, it is not immediately obvious who is actually interacting with whom. In a traditional unrecorded telephone conversation we can easily determine that two people are interacting directly with each other, without a go-between. The medium is like air in that it is just an open channel between one person's mouth and the other person's ear. Like an old ship's speaking tube, the conduit remains essentially unchanged after each word is spoken. By contrast, in the cases of the game characters and other abiotic proto-humems we discussed earlier, people are not just interacting with each other through artificial but *transparent* modes of communication. Rather, the proto-humems have become intermediaries of substance, which, once established, may persist indefinitely—even without the presence of their originators. Each of these proto-humems, in its own however rudimentary way, may possess person-like attributes such as long-term memory, knowledge, abilities, reputation, wealth, and charisma.

The interactions involving these entities are becoming increasingly complex and sophisticated. In some interactions, people are the dominant characters, but in others, people are more passive or not present at all. As humem capabilities advance, the behaviors that stem from their intrinsic makeup will become progressively difficult to distinguish from those that reflect the personalities of their human counterparts. As we become more integrated with our personal humems, and as our behaviors become more humem-like anyway, it may become increasingly pointless to even attempt such distinctions.

Alpha-Pairs

The affiliation between a person and the humem created from that person's EP is a fundamentally new kind of relationship that will consti-tute the basis of most human-humem interactions. To highlight this special and immutable bond, I label it the *alpha relationship* and refer to the union formed from the corresponding person and humem as the *alpha-pair*. This basic representation of the most intimate of the emergent person-humem associations underpins much of what we will discuss in this book.

For the sake of brevity, we can use the term *alpha* to designate the counterpart in the alpha relationship, whether it is the person or humem. This convention is similar to the use of the words "partner" and "spouse," i.e., the person's alpha is their corresponding humem, and humem's alpha is its corresponding person. In some cases, to explicitly differentiate between the members of the alpha-pair, we can use the terms *alpha-person* and *alpha-humem*.

When interacting with an alpha-pair, it may sometimes be difficult to detect where one member ends and the other begins. Often such a division will be unnecessary, and the person and humem members of the pair may be more aptly construed as complementary expressions of a single unit. This concept is not unlike how we often view a team, a couple, or other cohesive association with shared interests and assets. But here the bond is much stronger because the alpha-pair will almost always be inseparable. As we come to understand this relationship in more depth, we will discover that ultimately one will be unable to exist without the other.

Emulation of People Relationships

As reflections and extensions of individuals throughout global society, humems will be similar to people in terms of their diversity and complexity. To express this correspondence, we can refer to the collective of all humems and their institutions as *humemity*.

As with people's relationships, those of humems will be bound by conventions that regulate and direct their conduct. This applies in the soft sense, reflected by social and cultural norms, and also in the more formal realms, reflected by ethical and legal standards.

Mutually beneficial relationships between humems and the entities with which they interact will be crucial for humems' well-being and for their prospects for continuity and development. Especially in the early stages, humems will be crucially dependent on people and their institutions. If they relate and interact well with these other entities, they will be encouraged and nurtured. If not, they will be sanctioned and opposed. Thus, humems need to behave in a desirable and appropriate manner in order to thrive. As with all types of relationships, the prevailing norms and conventions will provide the guidelines for the interactions between and among humems, people, and their respective institutions.

For instance, the interplay between the partners in an alpha-pair will be characterized by an intimacy appropriate for this, the closest of relationships. Similarly, if a person communicates with a family member's alpha-humem, this interaction will resemble the closeness of the corresponding familial relationship. As a specific example, communications between a wife and her husband's alpha-humem would mirror those that occur between the wife and husband. This would be determined not only by generic conventions governing marriage relationships, but also by the specific nature of that particular marriage relationship.

Probably the best example of this idea is that of the parent-child relationship, as reflected by their respective alpha-humems. This is a complex and dynamic relationship that has biological, cultural, and legal underpinnings. Manifestations of this relationship vary widely among different countries and cultures. In addition, within a given culture, social norms and the corresponding legal constructs regulating this relationship are continually changing. Within a specific parent-child relationship, the phase of the relationship—based on the age of the embryo or child—also determines the nature of the interactions. For example, in most jurisdictions, the parent has extensive legal control and obligations regarding their young child, but most of these cease when the child reaches a certain age.

 Let's consider how the parent and child's respective humems express their relationship. Initially, a parent will establish a humem for the child, perhaps even before the child is born. Ultrasound images and data relating to the pregnancy are typical examples of the initial constituents of the child's alpha-humem. As we will see later, at the time of its creation the child's alpha may already contain extensive information, such as data relating to the child's ancestors, culture, and ethnic background. We can view this as a kind of *humem instinct*—a knowledge and aptitude possessed by the humem at its inception, or "birth."

The interaction between the parent and their child's alpha may be either direct or facilitated by the parent's alpha. As humems become more capable, the interaction via the respective alpha-humems will most likely become the natural approach.[25] This results in an interface between the parent's and child's alphas that mirrors the close parent-child relationship. For example, if the child's alpha notices that the child is behaving inappropriately—say, the equivalent of playing in the street in an online environment[26]—the child's alpha would be compelled to

---

[25] Already in the social network contexts, we are interacting predominantly via our proto-humems, which interact with each other. You log in to your account (or proto-humem), and from there you interact with other people's proto-humems; i.e., the proto-humems are the agents for our interactions. By contrast, reading a paper book copy of someone's biography is a form of more direct interaction with their proto-humem.

[26] For example, if the child attempts to transfer personal information to an online stranger, a capable child's alpha-humem should both block the communication and inform the parents.

share this information with the parent's alpha, which would, in turn, inform the parent.

When the child reaches the age of majority,[27] the formal parent-child relationship changes abruptly, which results in the child taking on the primary authority and responsibility for their own affairs. This shift is naturally reflected both in the relationships between people and their alphas, and in the inter-humem relations. While the parent's alpha-humem may have extensive control and oversight over the minor's alpha-humem, these too would end when the child becomes an adult.

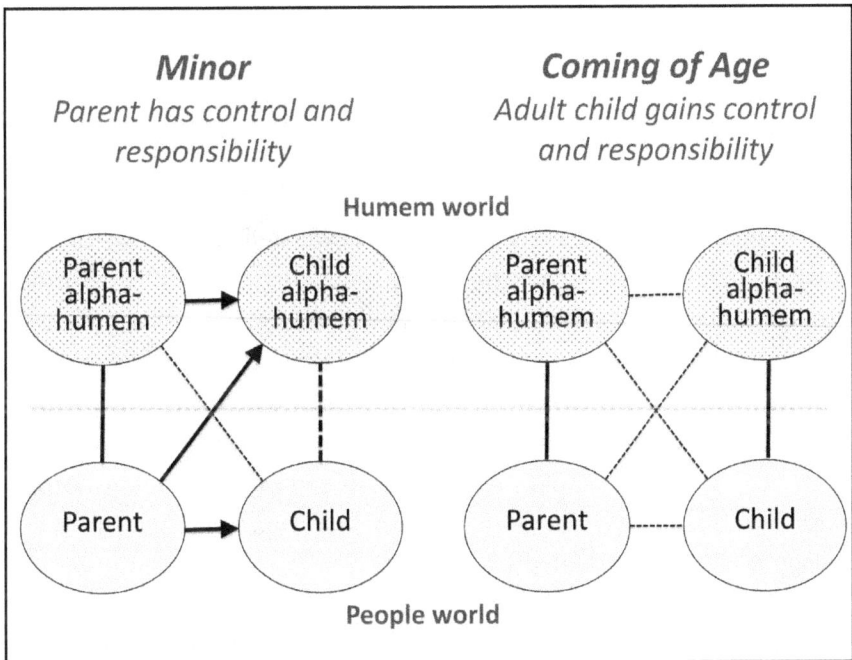

Figure 4: The parent-child relationship—control and responsibility changing as child matures.

Figure 4 depicts these parent-child dynamics. The solid lines signify a relationship of control and responsibility, with the arrows designating the asymmetrical parent-child relationship. Dashed lines indicate a relationship with less formal control, or none at all. The left diagram

---

[27] Also called "legal age" in some countries.

depicts these relationships while the child is a minor; the second diagram shows the connections after the child reaches the age of majority.

We see that since alpha-humems emulate and expand people's presence, the full span of inter-people connections will be reflected in inter-humem relationships. Moreover, a new dimension of people-humem relationships will interweave between humanity and humemity.

Companions—Lifelong and Beyond

In today's world the variety of human experience and behavior is vast. Some people spend their winter days on ice floes hunting seals, which they carry home on dog sleds in the dark. At night they dream of sea ice, dogs, and the return of the sun. Other people spend their days trading securities in stock exchanges. At night they dream of indexes, gains, and summer vacations.

The Inuit hunter uses the Internet to view satellite images of the ice pack when making hunting or fishing decisions, while the securities trader uses the Internet to research stock indexes when making buy or sell evaluations. Analogously, their alpha-humems will be adapted to their respective environments. In appearance and behavior the humems may be as distinct as their alpha-people. However, as is true for seemingly very different people, beneath these external distinctions, otherwise unrelated alpha-humems will also share many attributes—those things that distinguish them as humems. In the next chapter we will examine some of these distinctive humem traits in more depth.

It is possible to view one's personal humem as merely a tool that extends one's memory and cognitive abilities—perhaps just as an increasingly convenient method of managing one's life data. At first blush, this is how humems will likely seem to many observers. But for the vast majority of us, humems will ultimately be recognized as something far more anthropomorphic and expressive. As intensely social beings, we generally appreciate the company, guidance, and assistance of a like-minded companion. Of course, there are some things we may always prefer to do alone. Yet, these are apparently few and far between as evidenced by our proto-humems already accompanying us to places (for example, the bathroom) where we previously went alone. It seems that there are few situations where we wouldn't desire the company of a personalized and trusted escort. Let's briefly touch on a few examples to get a sense of what this might look like.

Children will be born in parallel with their alpha-humems. A baby's alpha can play a crucial role in helping to meet the baby's needs. As we'll see later, the humem will have many ways of sensing the world and sensing people. Thus, the humem can know the baby's detailed history directly, by accompanying[28] the baby throughout life. And it can also gain complementary knowledge indirectly, for example, by interacting with the parents' humems. From them, the baby's humem can learn things like aspects of the mother's medical status that could influence the baby's health. Data obtained by monitoring the baby and its environment, coupled with a comprehensive past knowledge, can allow the humem to determine the baby's health and feelings. Then the humem can conceivably speak for the baby. Parents often do not know if a baby is crying from hunger or from another pain. A baby's humem—equipped with the appropriate sensors and with access to the relevant data—may say to the parents, "I'm hungry!" or, "I have a fever!" Furthermore, with its precise memory and ever-growing computational power, the alpha can gain knowledge of things that the most attentive of human caregivers would never notice. It may, for example, discover a correlation between a particular mix of foods[29] and the baby's stomach-ache, and present those findings to the parents.

As we progress through life, the alpha-humem can function as a customized mentor who is aware of our abilities, strengths, weaknesses, needs, and motivations. The more humanized the humem agent becomes, the more effective it will be. Such a mentor may be able to

---

[28] We will gradually attain a sense of ways in which a humem can "accompany" their alpha-person.

[29] Rudimentary applications already exist which can identify food types. Thus, such a sensor adjacent to the baby could also detect what the baby is being fed.

ease our interactions with emerging technological ecosystems by adapting our experiences of the future to our persisting ancient emotions and impulses. We all have a deep need for the reassuring word of a parent, the guidance of a teacher, and the intimacy of a friend. These humem mentors can *complement* the human parents, teachers, and friends.

Let's imagine, for example, a schoolchild doing homework in the company of such a mentor. Now, some of us may cringe at the thought of such a scenario, citing a myriad of reasons why children should be doing their homework "by themselves," as children did in the past. But that era is over and will never return. Let's look instead at present reality and project it slightly forward. Already, many students, even while supposedly doing their homework "alone" in their rooms, are interacting freely with the world; they have open channels of chat or video linking them to friends, and are referencing the Web for their assignments. And this is just the beginning.

The humem can customize the learning experience for the child. Instead of wasting time or boring the child on topics they have already mastered, it can engage and challenge them with material that interests them at the appropriate level. In circumstances where children are already using electronic media to communicate with their teachers, the humem agent can facilitate more personal interactions by helping the teacher know when a child has difficulties and needs special attention. Similarly, many parents have little knowledge of their children's online lives, in large part because they are increasingly preoccupied with their own digital lives in separate rooms. A humem mentor can provide the parents with essential insight, and possibly induce them to become personally involved when necessary.

While we do not know in detail how all these interactions will play out, it is perfectly clear that there's no going back. We must embrace the future and make it better. We should also not confuse the current ungainly humem prototypes with the highly advanced personal humems that will exist in the future.

In general, the alpha-humem will have as wide a variety of manifestations and functions as its human counterparts. For some people, it will be a life-companion, an assistant, or an IT manager. For others it will be an alter-ego, a legacy bearer, or even a trust manager. For most of us, it will be many of these things, and more. An alpha-humem will at times be regarded as a separate but close entity, and in other circumstances, as a part or extension of the person. Commonly, the humem will be seen as having a parallel existence to its alpha, accompanying the person from conception, or birth, throughout their lifetime, and beyond.

The notion of alpha-humems persisting beyond the lifetimes of their alpha-people will be comforting and compelling to many people. Let's call these continuing humems, *ancestral humems*. In spiritual contexts, an ancestral humem may be perceived as a kind of soul—an incorporeal and continuing self. Later we will also see how ancestral humems can provide great practical value to individuals and society.

Choices and Decisions of the Alpha-Pair

As, at least partial, emulations of people, humems need a degree of decision-making ability in order to function and be useful. Often the alpha-pair will make decisions as a unit; but in some instances the humem may need to deliberate on its own. While I do not claim to know the specifics of how this will work, I will try to sketch a rudimentary picture of what this may look like—at least in the foreseeable future.

Let's first consider the mechanisms by which individual humems may be able to make decisions. As with humans, these will vary according to a vast array of environmental, cultural, legal, and other factors. But unlike people, the humem's inherent ability to decide and choose will also strongly depend on technological advances that will provide it with ever more sophisticated evaluative abilities.

The most elementary form of humem decision-making is perhaps equivalent to that observed in computers, which respond to input, or stimuli, using programmed "internal" algorithms. A more sophisticated variation of this kind of deliberation is when a system also utilizes its previous experience and "external" data—typically obtained via the

Internet or an intranet—to formulate its response to the stimulus. We are familiar with these kinds of dynamics from our daily interactions with an array of computing machines. An automated bank machine, for example, responds to a variety of input or stimuli provided by a user, and then "decides" to either dispense cash or refuse the request.

Here, one may ask whether the humem's decision-making capacity can extend fundamentally further. Could the result of a sufficiently sophisticated computer-run algorithm be comparable to a person's conscious decision? Similarly, is a person's conscious decision simply the result of a very complex algorithm that we have not yet fully deciphered? Clearly, the examination of these issues is beyond the scope of this book. I firmly assert, however, that humems will be able to make decisions at a sufficiently high level of human emulation to make this kind of functionality extremely useful and compelling. My confidence stems from the fact that such proto-humem behavior is already emerging in diverse areas.

For example, at the time of this writing, online applications, including some of the proto-humem examples presented earlier, are already making fairly good predictions about a person's preferences, desires, and decisions in certain circumstances. These operations, which are currently being performed primarily for marketing purposes, use only a small subset of one's total EP. When this kind of technology matures and is integrated into the personal, private humem (for its own benefit), it will be able to access and utilize one's full EP. Once this happens, humems will be much more capable of emulating a person's behavioral and decision-making processes.

In a long-term relationship involving two people—such as a sustained and close marriage—one person can often correctly predict their partner's preferences or decisions when presented with hypothetical situations. (Some television game shows have been based on this theme.) Similarly, once the alpha-humem has a record of the bulk of a person's abiotic-EP—including every incoming and outgoing mail and text message, every social media comment, perhaps even every word that the person ever said or heard—it will be able to predict with a high degree of accuracy how the person would respond in new situations.

Over time, the humem's aptitude in this role will most likely surpass that of any human partner because it will have a far more detailed memory of the person's past behavior and decisions in all facets of their life. Indeed, even the closest human partners are typically aware of only some of their spouse's choices. For example, while one may know her partner's personal-life decisions fairly well, she may not know much about his professional-life behavior or his demeanor in the company of his friends; she may have a good sense of his verbal communication style, but almost no knowledge of his written communications or online life. It is not uncommon to hear someone, on visiting their longtime spouse at work for the first time, claiming that their partner is "a completely different person at work." Not only can the omnipresent alpha-humem have much better awareness of all the facets of one's life, it can also recall much more than any person.

Furthermore, as it gains experience, the humem will also be able to evaluate its own abilities in this area. For example, from its past attempts at predicting the person's behavior in certain situations, and from subsequent comparisons with the person's actual actions or decisions, the humem can assess how successful it is at making such predictions. Thereafter, the humem can use this self-knowledge to fine-tune its future predictions.

In particular, building on a vast experience of what the person shares and with whom they share it, the humem will acquire a detailed knowledge of the person's privacy preferences. In this way, humems will grow with tact and privacy "built-in."

The humem can also superimpose its understanding of the alpha-person's past behaviors on a knowledge of current environments and changing cultural norms in order to make future assessments. As a simple illustration, let's consider a humem's ability to select clothes that correspond to its alpha's tastes. This skill could be employed to enable the humem to assist the person in shopping, and perhaps to allow the humem to emulate the person's appearance in the future. Let's imagine that this person almost invariably conforms to current mainstream fashion. The humem would become aware of this behavioral trait by comparing the person's past purchases with the fashions prevailing at the time, which the humem could ascertain by consulting relevant online resources. The humem could also determine, from its alpha's past behavior, other pertinent parameters, such as the person's preferred colors, sizing, and even their stylistic inclinations within the constraints of their conformity to fashion—say, their conservative versus flamboyant tendencies. By integrating the memory and understanding of the person's past preferences with the new cultural clothing norms, the humem can make recommendations that are likely to correspond with the person's current tastes.

In a similar manner, ancestral humems (alpha-humems of deceased people) will remain contemporary in many ways—not only in how they speak or appear, but also in their opinions on current affairs. People's views of the world are, by and large, composites of their inherent tendencies and the prevailing attitudes of the surrounding culture. Ancestral humems will have an intimate understanding of their alphas' innate characters, but will also be able to grasp the current mind-sets of society. A combination of these will allow them to represent their alpha-people as contemporary people.

Some people may object to this on the grounds that ancestral humems will not possibly be able to behave in the future *exactly* like their alpha-people would behave were they still alive. But this argument bears reflection: Is such an outcome, in which one behaves in the future exactly as one behaved in the past, to be expected or desired? While certain fundamentals of character or ideological leanings may remain fairly constant throughout most of one's life, a 50-year-old often makes very different decisions from a 20-year-old. We change and mature during our lifetimes. Thus, the notion of a developing and, to a degree, changing ancestral humem should not seem strange or objectionable. As I argued in chapter 1, any vivacious entity that exists in a dynamically developing environment needs to continuously adapt and gradually change in order to survive and thrive.

Let's now consider what we mean by alpha-pair decision-making. For the duration of a person's life we can think of alpha-pair decisions as those emanating from a couple. It is not unusual for one member of a couple to be more dominant during a given period or in certain circumstances. Likewise, in the alpha-pair's early years, while the person is a child, the humem may be more dominant in a number of areas. For example, the humem may control certain interactions with the external world. In today's context, this might entail shielding the child from certain inappropriate content from the Internet, or preventing the child from sharing personal information with strangers. In addition, it may communicate with the child's parents on certain matters—even against the child's wishes—such as notifying the parents if the child does not complete their homework or is cyber-bullying other kids. Some of the child's alpha-humem's modes of decision-making may be directed, or configured, by the child's parents. In particular, the alpha-humem may be able to inherit knowledge, like an understanding of the parents' expectations and culture from the parents' alpha-humems. We will explore these kinds of mechanisms in more detail in chapter 9, in the section "Reproduction, Sex, and Instinct."

Additionally, some aspects of the humem's decision-making may even be regulated by law. For example, notwithstanding the parent's instructions, in certain jurisdictions a child's humem may be legally obligated to shield the child from certain kinds of inappropriate digital content, such as graphic depictions of violence.

When the child comes of age, and typically for the duration of the greater part of their adult life, they would assume authority in most of the alpha-pair's affairs. A reversal of dominance would perhaps occur again in later life if the person became mentally incompetent, due to dementia or another infirmity. On the person's death, the now ancestral humem would assume full prominence in its expression of the alpha-pair.

This vision may give rise to concern regarding possible conflicts of interest between the person and their alpha-humem. Inter-couple relationships are not always harmonious. Human couples commonly experience conflicts of interest, such as those relating to the ownership of resources, the performance of work or duties, or incompatible expectations for their mutual future. Here, however, the analogy of the alpha-pair to a human couple breaks down somewhat. Because of their immutable interdependence, as well as the separation of resources between the human and humem domains (which we will discuss later), the members of the alpha-pair should have many fewer potential sources of contention.

As modern people, we often experience internal discord. Frequently this presents as a conflict between our current impulses and our more "sensible" or enduring interests. A familiar example is the desire for immediate gratification—say, by eating a delicious but unhealthy food—that contravenes our long-term health interests. We sometimes perceive such dilemmas as arguments between two separate parts of our minds. In such circumstances, the humem may need to take sides. I believe that while the humem will be naturally biased toward the more rational long-term self, it may well be able to mediate and help resolve such conflicts.

In the face of such a dilemma, we can easily imagine how the humem could impartially assist the opposing parts of the mind to vividly recall the results of comparable past decisions. As a rather indelicate illustration, the humem could determine how much alcohol the alpha-person just consumed. Based on past experience together with pertinent health factors like the person's recent food consumption and sleep, the humem may be able to reliably predict the effects of further alcohol intake—tomorrow's cost of tonight's cocktail. In the event of the humem foreseeing a negative outcome, it could graphically depict a past morning-after to help the now semi-lucid person to more clearly comprehend, and perhaps forestall, the likely consequences of another drink. Moreover, if the humem were so empowered, it could even be more assertive by temporary disabling the person's car keys.

Some of us, in comparing the scenarios depicted above with the capabilities of existing technologies, may be inclined to cynicism. But we must recognize that however superior our existing systems are to those of the past, they are still in their infancies. The digital information revolution has just begun and shows no signs of abating. We have to imagine what will happen when the individual humem's information content and its processing power increase beyond what they are now a thousandfold, a millionfold, and more. Barring a global catastrophe, such increases are inevitable and will occur in only a few decades. When these advances come to pass, functionalities that seem on the far edge of our current imaginings will become as commonplace as cars are today.

As with most of the examples that I present in this book, I do not intend to prescribe how humems should behave or interact with people in specific situations, but instead to provide a sense of the possibilities. As we progress in our study of humems, we will encounter many additional examples of how the alpha-pair makes choices, and how the allocation of control between the person and humem may change over time and circumstance.

The Changing Rules of Enduring Relationships

In the realm of people and their institutions, a relationship, once established, inherits a wide range of ethical, cultural, and legal norms. These include frameworks of rights and obligations that regulate the behavior of the participating parties. But unlike static contracts these conventions continuously develop and adjust, even while the underlying relationships may remain constant.

For example, in child-parent, husband-wife, and citizen-state relationships, amendments to the applicable legal frameworks often occur over timescales that are shorter than the duration of the associations. This situation is analogous to a sports contest in which the rules change mid-game.

Relationships may or may not be consensual. A marriage, at least in developed countries, is normally mutually consensual. The parent-child relationship is partially consensual, assuming that the parents willingly give rise to the offspring. The children, however, often to their subsequent dismay, have no say in the creation of this kinship. Similarly, the citizen-state relationship is most often a direct result of one being involuntarily born in a particular country.

One thing that all of these types of liaisons have in common is the unsettled situation in which the parties often find themselves. In entering into these relationships, they are bound by a kind of capricious contract, in which the terms are constantly being changed by someone else, and without their consent. For example, when you get married, even if you are the most judicious person possible you still only know approximately what obligations you are assuming. If you stay married long enough, some of the laws relating to marriage will change along the way. The same is true for the rules and conventions governing a citizen's relationship with a state: throughout the duration of the relationship, which commonly coincides with the span of one's life, these are constantly being amended. However, even in a democracy, the individual has an infinitesimal influence on the changes to or ordaining of specific conventions.

Why am I stating the obvious? Most of us intuitively understand why legal systems cannot be customized to individuals and why a changing environment requires an ongoing adaption of the rules. In the current humem context, the method of regulating humem interactions with others is not immediately obvious. Yet once we understand the similarities between the underlying natures of people and humems, and their dealings with each other, general guidelines emerge. Humems' emulations of human relationships will extend beyond casual exchanges to include formalized relationships, which are culturally and legally recognized. Thus, in the same way that people's relationships are constrained by cultural norms and regulated by the laws of governing bodies such as nation-states, so too do humem relationships need to be regulated by analogous humem systems.

The rate of change and the nature of the changes in humem relationships will resemble those of people. Moreover, due to the rapid development of the humem environment and the prospective longevity of humem relationships, the potential is even more acute for frequent changes of the regulatory systems to affect specific instances of enduring relationships. This applies to both the humem equivalents of interpersonal types of relationships—inter-humem and humem-person—and the humem counterparts of person-state relationships. An understanding of these dynamics is crucial for determining the suitable form of humem administration, which we'll examine later in chapter 7.

Non-Alpha Humems

At this point, it is worth noting that not all humems will be members of alpha-pairs. We can call these other humems *non-alpha humems*. To get an initial idea of what these may be, recall some of the proto-humem examples. For instance, we saw that social network elements exhibit the beginnings of humem behavior. These social network instances most commonly correspond to individual people and as such can appropriately be called proto-alpha-humems—initial expressions of alpha-humems.

By contrast, social network entities also sometimes represent an organization, company, imaginary character, or some other kind of non-person entity with an associated EP. These are early examples of non-alpha humems.

Though we will discuss non-alpha humems later in the book, for now we will focus on alpha-humems. Initially, at least, these will constitute the most interesting and complex humem manifestations. Once we achieve a better understanding of the alpha-humem, we will more easily conceptualize other types of humems. For brevity, whenever I use the term *humem* in the chapters that follow, I refer to alpha-humems. When we examine other kinds of humems, I will state so specifically.

CHAPTER 5

Distinctive Humem Characteristics

Now that we have a sense of humems' basic makeup and their modes of interacting with others, let's look at some of their distinguishing characteristics in more detail.

Anywhere—Distributed Locational Presence

People are tethered to time and place. Simply put, a person can only be in one place at one time. Humems, however, are not necessarily bound by such constraints.

Humans' location-dependence defines their behavior and shapes their environment. Partly because of this limitation, the world as we know it is divided into geographically delineated nation-states. This somewhat arbitrary division has, in turn, caused individuals' actions to be legislated based on their geographical location. In one place people may be allowed to freely perform an action that in another place could be punishable by death. Individuals sometimes leave a jurisdiction to evade the restrictions imposed by the "place." For similar reasons, many covet multiple citizenships. The ability to relocate and have agency in multiple dominions is invariably associated with greater individual freedom and safety. This world order is so deeply ingrained in our minds that it is hard to conceive of an alternative one in which physical location becomes irrelevant. Yet, as we increasingly use proto-humems to communicate and extend our personal presence, geographical boundaries lose relevance. Despite the fact that some governments still repress certain proto-humem expressions, especially in the social media realm, these kinds of EP largely transcend national boundaries. Exceptions, where they still exist, are often due less to geographic borders than to language and culture barriers. Moreover, the advantages of this extra-nationalization of humems are gradually being conferred on their alpha-people as well.

I am not contending that the nation-state division of the physical world is going away anytime soon. Nonetheless, major changes are starting to erode this world order. Children play online games together irrespective of their location—the primary hindrances are time-zone differences and meddlesome parents compelling them to sleep. We increasingly connect, share interests and opinions, and cooperate globally. Romantic partners have initial online interactions unimpeded by geographic divides. Sometimes they develop deep and lasting relationships that are only consummated much later through physical meetings. It's true that language and cultural obstacles still remain; but even these are steadily being reduced by the globalization of culture and the growing capabilities of the abiotic-EP—such as emerging real-time language translation applications. Consumers now bypass commercial importers and purchase a myriad of products directly from stores or individuals almost anywhere in the world. In these transactions, which are facilitated by means of reliable payment and shipping systems, we become increasingly oblivious to the location of the seller. Instead, we are more focused on the product, the price, and the seller's reputation, the latter of which can be readily determined through various expressions of their commercial EP.

The nature of work is similarly being transformed. Until quite recently, most people's work involved the manual manipulation of physical objects. Since the position of these objects was an essential factor in the outcome, their labor was inherently location dependent. Today, as the proportion of information technology occupations rises, many of us are able to conduct business in a progressively location-free manner. Of course, even computers are somewhere, but the specific placements of the machines and their users are becoming less and less relevant to the functions performed. In modern professions where we still "travel to work," it is largely for collaborative or managerial reasons—our need for eye contact—and not due to the whereabouts of the product. Even as the products are virtualized, we still are not. But as our proto-humems increasingly become the facilitators of our professional interactions, their advancing capabilities will eventually free us from the confines of place.

As we saw earlier, EP communication is becoming location and time independent. A number of individuals have long been able to influence the world in multiple places and at different times via their EPs—especially in the form of books. Historically, though, these exchanges have been mostly unidirectional and have given voice to only a small minority of the total population. Humems can do much more: they can facilitate location and time independence with multidirectional interactions for almost all modern people. We are already seeing the emergence of this behavior in some proto-humem manifestations, perhaps most noticeably in social networking applications. We can think of this as a *distributed presence* in both place and time.

In the traditional realms of human activity, globalization and its effects are progressively demonstrating that the old world order is ill suited to the future. For our modern forms of EP, and our humems in particular, these incongruities are even more conspicuous. Already, the meaning of humem location is becoming very indistinct. Over time, it will become almost meaningless. By their very nature, humems—like fish in the sea and dust in the sky—ultimately cannot be constrained by the old borders of geography.

This does not imply, however, that humems are completely exposed to the world. As we will see, alternative kinds of encapsulation are both necessary and achievable.

Anytime—Distributed Chronological Presence

Regardless of what our spouses and children sometimes say, as people, we generally do act our age. For humems, though, adherence to the present is mostly a matter of choice. We touched on this notion of humem *time flexibility* earlier in "Choices and Decisions of the Alpha-Pair." Let's expand this idea a little further.

Competent alpha-humems will always be up to date. Accordingly, they'll be able to simultaneously emulate and augment their alphas' characters. As the state of humemity progresses, these abilities will become ever more powerful and refined. Owing to their extensive memories, they'll also be able to represent people as they were in the past with increasing accuracy. Moreover, by applying this almost perfect knowledge of their alphas' life histories, the humems will be able to recognize tendencies in people's growth. Using these, together with the outcomes of other comparable people's lives, the humems should be capable of making increasingly valid predictions regarding their alphas' future characteristics.

This humem time adaptability will enable many novel applications. For example, a grandson may be able to interact with the ancestral humem of his deceased grandmother as she was at different ages. He may wish to experience her as a grandmother, during the latter part of her life, or instead, to interact with her as she was as a child. He may request, "Grandma, be my age!" Such interactions are possible and will be enthralling for many people. This does not mean, however, that all ancestral humems will be obligated to behave in this manner. Some ancestral grandmothers may prefer to keep things "nice and proper," and only interact with their grandchildren as mature grandmothers.

While most of us cannot consciously remember much about our first years of life, future alpha-humems will be able to recreate some of our earliest experiences. Thus, if we desire, they will provide insight into any stage of our personal past. They will notice trends in our development and discover causes and effects relating to our health and happiness. We will explore some of the far-reaching implications of this kind of enhanced self-awareness in chapter 10.

In addition to depicting the past as it was, a humem—like an "innovative historian"—will also be able to emulate the past as it could have been. Such "what if" scenarios could be invoked for entertainment, education, or for other uses. One may experience alternatives to one's real-life events—how things might have been if one's life had taken a different turn. Similar ideas have been widely explored in literature and in the performing arts—particularly in movies, where the use of sophisticated techniques for the manipulation of audio-visual senses makes such renditions especially effective. Humems, however, will facilitate much more personalized time-intermeshing applications.

Distributed Brains and Humem Dreaming

In present environments, humems reside in machines, such as computer systems, and interconnect via existing communication channels, such as the Internet. But unlike a person with a body and brain, a specific machine should not be regarded as the humem itself. Rather, the machine should be considered the transient abode or manifestation of a much more complex and enduring entity. For instance, if a particular computer or robot is to be somehow conceived as a sort of humem brain or body, then it should be seen as a transitory brain or body, and also, as just one part of a much larger distributed humem brain, body, and sensory system.

In our study of humems, we will gradually become used to the following recurring idea: while *conceptually*, or *organizationally*, the humem is a consolidation of parts into an individual, *physically*, the humem is a widely dispersed system. The conceptual and physical viewpoints are not contradictory at all. On the contrary, the distributed nature of the humem body, or physical existence, is the foundation of its resiliency, longevity, and versatility.

Comparable attributes can be found in existing entities, including institutions and people. An institution can be physically dispersed and still be organizationally, legally, and economically consolidated. This also holds true for personhood in its broader sense. A striking example of dissimilar individual spheres of influence and presence can be seen when comparing the stereotypical image of a homeless man with that of a celebrity, say an actress. The man's influence does not extend much beyond his bodily reach and the few possessions he carries on his back. Although he may wander extensively, at any given time he has a very localized existence. The famous actress, by contrast, has a much more distributed existence: a wide-ranging influence that extends far beyond her bodily presence. This includes her property and monetary assets, her imprint in various media, and her effects on other people's lives.

In addition to being distributed, humems' bodies and brains are also constantly changing and developing. Humems expand or transfer to new humem brain machines as these emerge. Like hermit crabs, they move to better abodes that facilitate their growth and development.[30] In modern contexts we can think of this humem relocation as *machine hopping*. Depending on the humems' affluence[31] and abilities, and those of their administrators, they adopt and take advantage of these more powerful and capable systems. In certain respects, these assemblies can be seen as generic brains that contain individualized and consolidated long-term memories. They are generic in the sense that standard machines and software are employed for different humem individuals, and they are constantly being upgraded and improved as technology progresses. Again, the consolidation of the humem memory should also be understood in an organizational sense: physically, it may and probably should be widely distributed for redundancy and survivability.

To a degree, the humem brain's properties and processes are comparable to those of people's brains. Moreover, the degree to which this comparison is valid will most likely increase over time. Although a human nervous system is encapsulated within a single body, it too is

---

[30] Of course, the hermit crab analogy is very partial: the shell does not contain the hermit crab's brains. In current terms, a more appropriate description of the humem growth is the migration of its body of data to more powerful computers with more sophisticated software.

[31] Humem affluence will be covered later in chapter 8.

often viewed as a decentralized system of semi-autonomous processing units lacking a clearly identifiable central control module. The compendium of these parts, nevertheless, results in a unity, which we call an individual. Analogously, although parts of the humem's memory and processing mechanisms may remain physically separate, their organizational consolidation into a humem will produce a recognizable individual.

A person's memory is unlike the fixed information contained within a copy of a book. It is much more dynamic, with living data undergoing continual replenishment and growth. Often, old data is lost or discarded, thereby making space for the new. Even though certain cornerstones may retain their form, they too are incessantly overgrown with recent experience. In this regard, the brain is more like a rainforest than a library. Similarly, the humem brain contains vivacious information that undergoes constant renewal and development over a core of persisting character. This kind of adaptability does not contradict the idea of long-term retention. On the contrary—as biology demonstrates so well—certain living information can persist for very long periods. By contrast, the static forms in museums and archives are, at best, subject to a delayed but inevitable decay.

As another humem-human comparison, humem background data processes such as defragmentation, indexing, de-duplication, and data correlations, are functionally reminiscent of some of the hypothesized information-processing purposes of sleep. Thus we may allegorically consider such internal data processing, even when the humem is not interacting with the world exterior to itself, as a type of humem sleep and dreaming.

Distributed Sensing and Body Awareness

Humems will be able to experience the physical world. From a sensory perspective they will emulate us in many ways. Furthermore, they will often surpass our abilities. In particular, the humem's sensory system, like other facets of its existence, can be in many places at once—it is a distributed sensing system.

In many areas, machines already sense much better than we do. Imaging and sound recording technologies allow the capture of visual and auditory sensory data with much greater precision than that of the human sensory system. Machines can detect the full span of the electromagnetic spectrum, while people are only capable of sensing a tiny segment of the total. Likewise, machines can sense temperature, humidity, direction, and location with much better precision than we can. In a few areas, like smell, we still do better. Yet, these kinds of sensing machines have only recently emerged and considerable improvements are inevitable. In short, it is clear that humems will emulate and probably exceed most of our sensory abilities in the future.

In many respects, people still maintain far superior abilities relating to the *processing* of the sensory data, such as pattern recognition in visual or auditory records. But here too, machines are making steady progress.

The sensors in our mobile electronic devices—microphones, cameras, location-determining systems, sports and health monitors, and others—are conspicuous examples of our proto-alpha-humems' sensory organs. It is most likely that with decreasing costs and more advanced applications, a myriad of sensors of various kinds will soon be found in and on our bodies, and in our surroundings. And all these will be able to contribute to the humem's sensory input. With the help of this hoard of sensors, humems will be able to achieve a high level of awareness of the environment and of the bodies of their alphas.

I would argue, however, that humems' sensory systems should be viewed even more broadly as all the apparatuses through which they absorb information from the outside world. There also exist more indirect and subtler humem sensory channels. For example, when a person has a blood test, their humem may receive the results via an electronic communication. As long as the humem can access this information, we can consider this to be humem sensing. Here, we can simply think of the medical laboratory as an extended and transitory humem sensory organ. In other words, a blood sensor on the person's body (if or when such a device exists), or a machine in a distant lab, may provide comparable results. Similarly, sensing environmental factors such as the ambient temperature of a person's surroundings can also be performed directly, via a temperature sensor worn or carried by the person, or inferred indirectly, by using the person's location information (as determined by their mobile device, for example) coupled with the measurements of a building's environmental monitor or a local weather station.

We see that, unlike a physical person, whose senses are devoted or restricted to their body, the humem can utilize sensors that are not necessarily dedicated to its sole use. Although the individual humem is a consolidation of purpose and character, its sensory system—like its body and brain—is physically dispersed in location and time.

Going Places—Mobile Humems

Even though the humem as a whole is not localized or constrained to a physical place, we can think of the humem as "being" or "going" somewhere in a sensory manner.

We can imagine the humem escorting its alpha by employing the alpha's mobile device as its sensory system for a sense of place and direction, a sense of cadence, a sense of time, a sense of sound, a sense of sight, and so on. However, at the same time, the same alpha-humem could have other comparable memories of places on the other side of the world, while similarly accompanying the alpha's child on a school field trip (by means of an application on the child's mobile device, for example). Likewise, ancestral humems may be able to accompany their descendants, if desired.

Accordingly, a humem can experience being in a specific place at a specific time—it can have memories of a geographical, time-dependent event. But the humem can also be in many places at the same time—it can have multiple memories of concurrent, but geographically separate events. Moreover, it can have memories of other concurrent happenings that are unassociated with any location, such as the reception of a digital message or an event relating to a social media proto-humem. Still, as we have previously observed, the humem's memory and brain processes, and other material assets, or parts of the humem body, are geographically dispersed. So while we can say, in a very specific way, that the humem can *go* places, it does not make much sense to say that the humem, as a whole, is ever *in* any single place.

Substrate-Independent Humems

Once they become more expressive, there may be a tendency to think of humems as people-like emulations that exist on digital media and are animated by electronic machines. This view is short sighted—both backward, into history, and forward, into the future.

In the same way that Cleopatra's EP lived on various types of substrates, or media, over the ages (bio-EP in people's brains, writings on papyrus, statues of stone, metal coins, print in paper books, oil paintings, films, and digital media), humems are media-independent beings.

At any point in history, we tend to construe the current state of the art as the final word. But we are always mistaken. We have little knowledge of the material nature or mechanisms of the future EP carriers.

It is therefore clear that, wherever possible, we should establish the humem entities to be independent of any specific media. Where this is unattainable, contingencies for future substrate changes should be kept at the forefront of our long-term planning.

The Humem-Self

In people, there is currently no established physical counterpart to the notion of the "self." As mentioned earlier, the brain seems to behave like a decentralized system, without anything resembling a control center that we could identify as the core of the person, or the self.

For most of us, this fact does not detract from our identity as individuals. The notion of self is one of those things that we just know, or assume for all practical purposes, without necessarily having any understanding of the underlying mechanisms. Furthermore, the self is widely recognized in our cultural and legal systems.

As humems grow in complexity and capability, and as they gain increasing levels of agency and individuality, I postulate that this progress will culminate in the ultimate expression of a person-like being—the emergence of the *humem-self*.

While it may initially be possible to tabulate the discrete components constituting a humem, as complexity increases, a humem-self may emerge whose underlying workings may not be fully understood. The difficulty of the reductionist problem for humems may approach or surpass that of people—or more likely, it may be inseparable from that of people. However, as with the human self, this lack of understanding of the fundamental mechanisms should not prevent a broad pragmatic adoption of the idea. As I have repeatedly suggested, the recognition of the humem-individual will probably be completely intuitive in most situations. Furthermore, based on the human precedent, it is likely that the practical application of the concept of the humem-self will also extend to more formal domains such as those of economics and jurisprudence.

Part III

ESTABLISHMENT

CHAPTER 6

RIGHTS AND CITIZENSHIP

Presently, proto-humems exist in what amounts to anarchic wilder-nesses–digital jungles—embedded within organized human systems. We can view such a wilderness, or ecosystem, as a nature preserve that is, as a whole, administered by people, but whose inhabitant proto-humems are not recognized as individuals. They have no personal rights or protections and are therefore subject to the "law of the jungle." It follows that although the humem ecosystem in its entirety may have good long-term prospects, its residents—as individuals—do not.

As we have begun to see, humems will have a multitude of expres-sions. In the personal realm they will emulate and expand social interactions. The better they do this, the more we will desire them. They will act as our agents, enhancing our communication with the world. Some people will view humems as an extension of their personal-ities—an inseparable part of themselves; others will look upon them as a kind of offspring. In spiritual and religious domains they will facilitate more tangible representations or extensions of notions such as the soul, the spirit, and ancestral presence. As a result of these personal affini-ties, we will come to desire for our humems many of the same things we require for ourselves, including agency, freedom, welfare, security, and even economic opportunity. As we will see, without such rights and protections, humems will be unable to fulfill many of the functions that we will expect of them.

Since proto-humems are becoming increasingly similar to us, serve our needs, and are sometimes seen to be under our control and ownership, some people may be inclined to think of them as a kind of slave. In traditional master-slave relationships, slave owners have been faced with a dilemma: whether to promote their slaves' education, thereby making them more useful, or to constrain their development, thereby inhibiting them from pursuing freedom. No such quandary need exist for humems; increased knowledge, capabilities, and other forms of empowerment will largely make them more useful and appealing without diminishing their affiliation with their alpha-people.

I maintain that an essential component of such empowerment is the endowment of humems with rights within the framework of a citizenship similar to those we find in developed countries. In this chapter we will start exploring what these rights and citizenships might look like. I will also argue that alternatives, in which humems have little freedom or agency, result in multiple shortcomings. As humems emerge and grow in their capabilities, preserving their existing status will result in stunted and underutilized humems, or worse, humems that act to the detriment of others.

Citizenship implies governance within a state—capable and empowered humems will need systems that ensure their well-being and regulate their behavior. We will see that due to the growing similarity between humems and people, humems require a state that, in many respects, is similar to traditional states but that also has some fundamental differences and special adaptations to humem needs. I will argue that, at least for the foreseeable future, our existing state institutions are unsuitable for humem administration and consequently humems require a state that is separate from, and eventually independent of, the alpha-people's nation-states.

I readily acknowledge that the proposition of a state for humems may, at a first glance, seem unnecessary or extreme. However, I hope to convince you that this approach is absolutely necessary and the natural way of organizing humems and the humem ecosystem. In accordance with this vision, it is appropriate to call the humem governing institution the *humem-government*, and the broader environment that encompasses the government, the humems, and the infrastructures on which they exist, the *humem-state*.

In this chapter we'll explore why humems require rights and citizenships, and in the next chapter we'll examine the characteristics and path to establishment of the humem-state.

Cleo Dethroned—Classical Deprived Humems

In chapter 1, we surveyed the resiliency, longevity, and development of Queen Cleopatra's EP. We also saw that most people's EPs are already becoming similar to those of celebrities in many ways. It is reasonable to ask then whether it is both *probable* and *desirable* that ordinary persons' EPs will persist in the future as those of celebs have survived in the past. Along the same lines, we can ask whether an organized and formal establishment of empowered humems is really necessary for the long-term survival of the EPs of the masses.

First, let's consider whether it is *probable*: Will common people's EPs persist in the future even without the establishment of humems? In the past, commoners' EPs did not survive, so there is no precedent to point to. Yet, commoners in the past did not accumulate the vast amounts of abiotic-EP that ordinary persons have today, and they did not have the means for its broad dissemination. In addition, it now seems likely that a large part of this present form of abiotic-EP may remain for extended periods. For example, deceased individuals are already leaving massive digital "footprints" while the rapidly declining costs of storage and the institution of digital archives suggest that these digital EPs may endure indefinitely. Thus, in this case, the past is probably not a good indication of the future.

Let's end this line of inquiry promptly by admitting that this question is simply not answerable at this time. With the current trajectory of EP growth and behavior, no one can predict with any certainty how much of the average person's personal EP will survive into the future. It may be fifty or a hundred years before we'll have initial answers to this question. So with this unknown identified, it seems reasonable to establish the following working assumption: While no one can foresee how much personal EP will persist by default, we can assume that the establishment of humems, with dedicated organizations and resources, would significantly improve the prospects for the continuance of ordinary people's EPs.

Now we can go on to examine the more pertinent, and in my opinion, more easily answered, part of our original question: Is it *desirable* for our personal EP to emulate the achievements of Cleo's and other celebs' EPs? Is this what we really want?

Although Cleo's EP is massive and well established, it nonetheless suffers from some fundamental deficiencies. For one, it is fragmented and dispersed like most current and past EPs. Because it declined significantly in the past, a large portion of Cleo's present EP has diverged considerably from what it was during her life—that is, much of her EP has been reinvented and thus is less authentic. As we have seen, adaptation and change are necessary for the survival of personal EP. However, the authenticity and stability of the core EP character are also of essential worth; without them the notions of continuation and persistence are meaningless.

But, perhaps the most significant deficiency in Cleo's EP is that whatever remains of its more authentic parts is derived almost entirely from her public EP. Virtually nothing of her genuine personal character has survived. Intimate expressions of her personality, as depicted in scenes in movies or books or plays, have all been recreated by other people and have a dubious resemblance to her original nature. Here we should note that this situation is less a consequence of the extended lapse of time since Cleo lived than it is a result of the manner in which Cleo's EP was first created. From the outset, Cleo's abiotic-EP almost exclusively reflected her public presence. There were no substantial recordings of her personal conversations with family, or private journals depicting her intimate thoughts; that's just not how things were done back then.

Even if a part of a celebrity's intimate EP were to survive, it would eventually face the loss of most of the associated rights that it enjoyed during the person's life. Today, in many developed jurisdictions, a person's private EP has a number of legal protections during their lifetime, such as those afforded by laws against defamation. But once the person dies, many of these defenses crumble. In most legal systems, for example, there are no protections for the reputation of the deceased.[32] Cleo has no remaining rights at all. She has no claims to privacy, honor, or respect. She has no agency relating to the perpetuation of her legacy, or any copyright on her character. At this point, many may ask: How can this be any different? Why should this be any different?

Let's first consider the implications of ongoing privacy or the lack thereof. We have private thoughts and memories, many of which we do not want to share yet would still want to retain even if presented with the alternative of somehow erasing them. Cherished but private memories might be valuable both intrinsically and in the ways that they influence and enrich our outward characters. Currently, abiotic extensions of these facets of one's mind—such as a diary—are inherently at risk of either destruction or exposure. Faced with this uncertainty, many people opt for their deletion. And since people know that future circumstances might prevent them from erasing these traces, they often destroy them prematurely. The alternative is the fear of someday losing control—epitomized by the vision of lying helplessly on one's deathbed agonizing over the future exposure of personal, family, or legal secrets.

We also possess less private kinds of information, which are not intended for public disclosure but rather to be shared with a limited group of people. Once this material becomes a part of one's abiotic-EP, it becomes difficult to manage. Most commonly, the family, friends, or colleagues of a deceased person de facto inherit the responsibility for supervising this private information. In practice, however, the subsequent regulation of this EP's distribution is often impractical and is not handled in accordance with the wishes of the originator. The more

---

[32] This is not always the case. For example, previously in California, under Penal Code (1872)—sections 248 on slander, " tending to blacken the memory of one who is dead," was illegal.

dispersed the EP, the more chaotic and unsuccessful these endeavors become. Here too, the usual result is either the loss or, conversely, the broad distribution of these data.

Some may flippantly ask: Who cares about their privacy and reputation after they die? Well, it turns out that many people do passionately care about their legacies and how they and their families will be perceived after they are gone. This is why many people write memoirs and autobiographies. Moreover, the loss of privacy of the dead can sometimes be very detrimental to the living. The rapidly growing portion of our EPs being recorded as abiotic-EP and having overlaps with the EPs of others is making the privacy problem much more serious than it was in the past.

Returning to the question posed at the start of this section, we can conclude that while most people would want the bulk of their EPs to persist indefinitely, they would not want the EPs to lose their personal quality or to have their full content exposed to the world. They would not wish to have their EPs emulate Cleo's denuded and impersonal EP. Were it possible, most people would opt for the perpetuation of both their privacy and the protections afforded to their reputations.

The establishment of humemity will greatly increase the likelihood of the survival of a personal EP. But, just as important, humemity will ensure the perpetuation of the EP's true character. In this domain too, we find that ordinary persons' EPs as carried by humems can attain superior existences to those of the monarchs of old.

Humem Rights

Universal human rights and citizenship rights are often accepted as the definitive frameworks for ensuring people's well-being and protection from gross injustice. I assert that humems, as an extension of people, require similar rights for similar reasons. We will discover that humem rights are inextricably linked to human rights, and the attainment of fundamental rights by humems will supplement and enrich those of their human counterparts. In particular, humems' gradual preparation for emancipation and independence is absolutely essential for them to achieve our aspirations for their continuance beyond our lifetimes. Moreover, we will see that since humems exist in material domains that are mostly separate from those of people, humems' essential rights need not detract from those of people.[33]

Arguably the best-known decree in this field is the Universal Declaration of Human Rights (UDHR), which was adopted by the United Nations General Assembly in 1948. I suggest that most of its thirty articles apply, in one way or another, to humems. I will refer to some of these UDHR articles as a backdrop to the following discussion. (The full version of the UDHR is readily accessible on the Internet and elsewhere.)

Of course, there is no unanimous agreement on human rights. The rights and their definitions continue to develop and change, and we see significant variation in different cultures. It would therefore be imprudent to assert that any one document could establish a broadly applicable charter for humem rights. Still, it's valuable to examine some of these fundamental ideas and consider how they may relate to humems.

The Right to Rights

The UDHR's first articles clearly define the entity deserving of human rights as any living person.[34] Thus, in principle, its applicability criteria are simple. Nevertheless, in practice, in the past and still in many current jurisdictions, the actual acceptance criteria for one deserving of such rights are much less obvious. In many states in the world today,

---

[33] Historically, resistance to the allotment of new rights to people or animals has invariably been centered on conflicts of material interests. We will discuss this issue in some detail in chapter 11.

[34] UDHR-Article 1: All human beings are born free and equal in dignity and rights. They are endowed with reason and conscience and should act towards one another in a spirit of brotherhood.

UDHR-Article 2: Everyone is entitled to all the rights and freedoms set forth in this Declaration, without distinction of any kind, such as race, color, sex, language, religion, political or other opinion, national or social origin, property, birth or other status. Furthermore, no distinction shall be made on the basis of the political, jurisdictional or international status of the country or territory to which a person belongs, whether it be independent, trust, non-self-governing or under any other limitation of sovereignty.

someone who was born in another country or is of a different religion or ethnicity from the ruling class can never attain equal rights or citizenship. Not so long ago, even in some of the most developed places in the world, women or people of certain races were lawfully denied full rights. In some societies, this is still the case today. The point here is that despite universal guidelines such as the UDHR, human rights are not yet implemented globally. We see that the determination of eligibility for human rights is an evolving and dynamic process. It is almost universally true that, in most places and cultures, human behavior that seems inconceivable or repulsive today was previously the norm. These transformations often occur even while underlying ethical directives—as established, for example, by the prevailing religious doctrines—remain nearly constant.

Similarly, the future criteria for humems' eligibility for humem rights are not obvious. For one thing, it is not altogether clear what may comprise a "complete" humem deserving of such rights, as opposed, for example, to a proto-humem or some EP fragments. There is no simple criterion equivalent to the relatively clear-cut biological definition of a living person. (Comparably, for people too, there is an ongoing debate on the rights of embryos and fetuses.)

Even so, this does not mean that it will be impossible to make such distinctions in the future. We can infer from the history of people and states that the recognition and definitions of humems will change dramatically over time and will vary significantly between jurisdictions. In practice, specific humem administrations—the counterparts of state governments—will probably lead the way in defining humem rights. Such has been the progress on almost all fundamental human rights: individual states establish new norms, others follow, and the international conventions come later. To the extent that such systems are beneficial to humems and their human relatives, and provided that they do not overly infringe on the sensitivities and interests of others, many such environments will emerge and endure.

Since humems are increasingly acquiring people-like qualities, in order to envision how humem rights and our perceptions of humems' *right to rights* may develop in the future, it is worthwhile considering how societies have amended their view of the rights of people-like entities in the past. In the current context, "people-like" entities are defined as ones that are perceived to possess a significant number of human attributes, but which the associated society does not regard as being equivalent to people. For instance, because many people treat their dogs almost like children, it is reasonable to say that they ascribe a substantial level of "people-likeness" to these creatures. Animals like frogs or fish are generally deemed to have fewer such attributes and, in the eyes of most people, bacteria and plants have virtually none. On the other end of the scale, higher animals like apes and dolphins are increasingly humanized—especially in depictions in modern media— which is, of course, the primary medium through which most people experience them.

These changing perceptions have tangible outcomes that influence our interactions with these beings. In biology classes, contemporary first-world school kids have few qualms about killing a million bacteria or some fruit flies. Most do not have an ethical aversion to dissecting a frog, but may feel some unease in harming a mouse. For similar purposes, a dog or cat is typically out of bounds, and most children would not even dream of injuring an ape or dolphin. Clearly, many of these sensitivities are very new and correlate closely with the degree to which we regard these animals as being similar to ourselves. These feelings are in turn reflected in our legal systems and in the institution of new rights and protections for these beings. In medical research, for instance, certain practices involving animals that were commonplace fifty years ago are now deemed unethical, illegal, or both. Currently, in effect, a pet dog in a developed country often enjoys greater state protections, better nutrition, and more advanced medical care than many people in impoverished states.

However repulsive or implausible it may now seem, in most places at some time in the past, slavery existed and slaves were viewed in a similar way to how we presently view higher animals—possessing certain human attributes without being fully human. Significantly, this was not just the position of a few deviant individuals within those societies but rather the worldview of the majority of the dominant class of people. By the laws of those countries, slaves were often not even second-class citizens; instead, for all intents and purposes, they were considered property. In fact, in some jurisdictions, slaves had fewer rights two centuries ago than some house pets have today.

I believe that this overview of societies' changing attitudes to people-like entities provides insight into probable future sentiments toward humems. A view of the historical transformations in those things that were once ordinary and now seem inconceivable, and vice versa, can recalibrate our sense of the possibilities in this area. Specifically, for the sake of our analysis, I think it is useful to compare the formal status of humems today to that of slaves in the past. Some may find the comparison of human slaves to supposedly "artificial" entities objectionable or in bad taste. This parallel is not, however, intended in any way to diminish the legacy of suffering by slaves over the centuries; rather, it is meant to illustrate how cultural and legal conventions change over time in regard to how societies relate to what they see as merely people-like entities and their rights. Also, when we begin to view humems more as an extension of people and not as something separate or artificial, this analogy may become less offensive and more natural to those who currently dislike it.

As we proceed in our examination, I'll use this parallel to elucidate some of the deficiencies of the current humem systems—how humems' current status of property impedes their development and diminishes their full expression. I also believe that this analogy provides some guidance as to the means of humems' emancipation and endowment with rights.

The Right to Come into Being and to Exist in Perpetuity

In article 3, the UDHR declares the right to life, liberty, and security of person.[35] We may perhaps also call this the freedom from existential intimidation. Notably, it doesn't specifically state the right to be born, although birth is partially covered in a separate article relating to the right of a couple to create a family. So, from an individual's standpoint,

---

[35] UDHR Article 3: Everyone has the right to life, liberty and security of person.

birth itself is not clearly stated as a right; the composers of the UDHR seem to be saying that regardless of how one came to be, these are their basic rights.

Deliberating the right to be created or born places us in the middle of some thorny ethical and religious controversies, in particular, that between the "pro-choice" and "pro-life" positions. The ongoing lack of consensus in this area is presumably the reason why it has been avoided in the UDHR. Recent progress in artificial interventions in conception, pregnancy, and birth are challenging the validity and sufficiency of most previously accepted norms and legal frameworks. The deliberations regarding an individual person's right to come into existence are becoming ever more complicated and are still far from being resolved.

On the other end of the lifeline, for now at least, matters are more clear-cut: it is presumed that all people eventually die. Thus there is no major concern about perpetual individual rights.[36] In most societies and jurisdictions, people's rights cease on the occasion of their death or shortly thereafter. Over the centuries, dedicated legislation has prevented the continuity of people's control or ownership beyond their lifetimes.

For humems, however, the situation is markedly different in both of these areas. As we previously saw, the processes relating to humems' coming into existence are not clearly defined. Therefore, the right to come into existence also needs clarification. There are many ways in which this might play out. It may be as simple as saying, if it can, it does! That is, if a humem can come into existence and there is a jurisdiction that will accommodate it, then it will be recognized and endowed with humem rights. In the initial stages, where existing nation-state jurisdictions may regulate the way in which people create and allocate resources to humems, more stringent criteria might apply. In general,

---

[36] There have been some marginal forays into this area, such as those initiated by the proponents of Cryonics (the freezing of bodies for future reinstatement). However, there have not been any major frontal attacks on legislation in this regard. Rather, the supporters of these endeavors have circumvented current statutes to achieve their interim goals.

humems do not have any absolute cause to terminate, or die. Indeed, they have many reasons, and the ability, to endure indefinitely. Thus, unlike people, the right to exist in perpetuity[37] is primarily a practical issue for humems. Jurisdictions hosting humems will need to provide these protections; if not, others eventually will.

The Right to Termination and the Right to Bequeathal

Having just discussed the right to exist in perpetuity, it may seem contradictory to propose a right to termination, or death, of the humem. However, humem continuance should probably be a right and a choice of the alpha-pair, but not an obligation.[38]

I think that it is unlikely that the right to termination will be invoked frequently, but in deference to the sensitivities of the people who establish humems and give rise to humemity, I believe that this choice should exist. For any who may be disinclined to perpetuity, we will see that the humem system is also able to facilitate "graceful exits" much more effectively than those possible with a fragmented EP. We will discuss this idea in chapter 11, in the section, "When I'm Gone, Erase It All! The Ashes-to-Ashes Aspiration."

This right should include the freedom of choice for the timing of such a closure. That is, it should not necessarily have to coincide with significant life events such as the death of the alpha-person. The process should also include the possibility of occurring in stages—a gradual and organized dissolution as desired by the alpha-pair.[39]

---

[37] I introduced the notion of "perpetuity" in chapter 1 where I argued that not only has Queen Cleopatra's EP survived and flourished for the last two thousand years, but it is actually hard to conceive how it could not continue in the future. There is no discernable upper bound to its ability to persist, and there are many tangible reasons to believe that it will continue to do so. This is broadly how perpetuity should be understood in its current use.

[38] It is my opinion that perpetuity should be optional. I predict, however, that when the humems resemble people more closely, there will be those who will disagree and argue that it is unethical to allow a humem to be erased.

[39] This has pertinent implications today. In some cases this may be regarded as an organized and structured data destruction or data access bequeathal.

The right to termination implies a right to bequeathal of the humem's associated assets. In people jurisdictions, estate law is often a complex and highly regulated affair. With humems, beneficiaries may include both people and other humems. Eventually, in this area, the complexity of humem systems will probably be at least as intricate as that of the corresponding people systems. However, for these and other humem rights, early solutions will have to be considerably simpler.

The Right to Autonomy

Conditions of slavery and servitude have often originated in the capture of free and able people. Yet, there is another frequent path into servitude that occurs when the conditions of the servitude, dire as they may be, are nevertheless preferable to the alternatives. Often this simply means that servitude with food and shelter is still better than starvation and death.

However less dramatic this may appear in the humem context, many proto-humems are in a similar situation today; they can either subsist under the control and ownership of corporations and other masters, or not exist at all. We will study some examples that exemplify these circumstances later in this chapter.

At a first glance, the present state of affairs may seem somewhat natural due to humems' currently limited capabilities. For now, they are incapable of independence. So why not wait for humems to mature and address these issues later? First, this will be a self-fulfilling prophecy. If we treat humems as slaves, as such they will remain for a much longer time. Second, when they eventually are emancipated, they will still have the character of slaves. The mentality of one born into servitude can become entrenched and persist long after becoming free.

In considering the best environment for alpha-humems, who are temporarily less capable but no less cherished individuals, we should think instead of how we raise our children. Children's development is a gradual process: as they gain capabilities, they are granted greater levels of autonomy. While childhood is significant in itself, we seldom

lose sight of the destination—the time when they will be fully independent. Like a child, the alpha-humem should be afforded the opportunity to develop in parallel with its alpha. Unrestricted by artificial
barriers and starting to learn as soon as there is anything that it could
and should learn, the molding of the humem's core character should
start as early as possible. This will maximize its capacity to function as
an emulation and expansion of its alpha-person.

Accordingly, early humems can be seen as existing in a state of
humem childhood and being cared for by *guardians*. (The primary
guardian will typically be the humem-government. Depending on the
guardianship functions performed and multiple other factors, such as
whether the humem is an ancestral humem, the alpha-people may also
have important guardianship roles.) We will find that this life-stage
metaphor remains consistent over multiple areas of humem progress
and will greatly aid our intuition of humems' developmental years.

We must be careful, however, not to confuse this humem childhood,
or childhood *of* humemity, with the humem of a child in its individual
development, or what we may call a childhood *in* humemity. While
there may be certain overlaps in their properties, these two concepts
are fundamentally different. In the future, when humemity has
matured, the childhood of humemity will be over. But, even then, a
child's alpha-humem may be purposely constrained in certain behaviors
as an emulation of the child, and not because humems in general are
incapable of "adult" behavior.

We can now rephrase humems' right to autonomy in a more
nuanced manner. Humems are only just emerging and not yet capable
of advanced autonomy. Thus, in this interim stage, the right to autonomy should be understood as the right to *develop toward autonomy,* as is
typical of a well-structured childhood.

Correspondingly, humem institutions should be designed and built
while taking these objectives into consideration. Also, no obstacles
should be established that would preclude the eventual achievement of
autonomy by humems when they become sufficiently capable. In
conclusion, while the path to humem autonomy may pass through
childhood, it should never entail servitude.

The Right to Legal Justice and Due Process

A number of UDHR rights relate to justice, due process, and equality before the law. Once an entity has agency and is permitted to own assets, it inevitably also becomes a participant in the legal system.

Some kinds of non-person entities already possess legal rights and liabilities. States, corporations, and estates are often regarded as non-human legal bodies with various forms of rights and accountability. Thus, the idea of humems becoming recognized as liable and responsible legal entities is not as extraordinary as it first may appear.

One of the most perverted constructs of the institution of slavery was the notion that a slave was not considered a person in the eyes of the law except under criminal accusation, in which case they could be brought before a court of law and tried—more or less—as a citizen. A similar situation—in which, on the one hand, a humem has no right to legal protection, but on the other, is liable for any detrimental effects that it may inflict on other parties—would pose a grave risk to the humem's welfare.

In early stages, the humem governing body (with perhaps the involvement of the alpha-person), in fulfilling its role of a guardian during the humem childhood, will be responsible for regulating humems' interactions with the world. Guardianship, we know, is profoundly different in character than ownership. An owner is primarily economically motivated, whereas an appropriately appointed guardian's core interest should be the ward's welfare and development.

Since humems will initially be in a state of childhood, they will not be legally liable. During this phase, like a parent, the guardian will assume both control and responsibility. This humem childhood must, however, be understood as a temporary condition, the duration of which is not currently clear. Notably, the humem, with its outstanding prospects for longevity, may not necessarily be in any great hurry to end its protected childhood. In the interim, it is the guardian's duty to prepare the humem for a more independent future. Once this emulation of childhood is established, a safe and unhindered humem growth can commence; then, there will be ample time for the judicial, regulatory, and technological frameworks, required for greater humem independence and responsibility, to be gradually developed.

The Right to Enduring Privacy

Earlier in this chapter, we saw how Cleopatra's and other celebs' EPs, as expansive and long-lasting as they may be, are devoid of privacy, and how this detracts from their authentic characters. Let's now expand this discussion and examine the need for ongoing EP privacy from the humem perspective.

We may never share the deeply personal parts of our minds with others, yet they are an integral part of our characters. Moreover, internal memories and dispositions, while private, have many external, or public, manifestations. Springing from hidden sources, the inner self is filtered as it percolates through layers of discretion—ultimately surfacing as one's outward character.

For most people, the degree to which they share or expose their innermost thoughts and feelings is constantly shifting, influenced by many factors: some interior, like one's age and inherent tendency to openness, and others exterior, like the influence of specific relationships or changing cultural norms. Today, as in the past, people with contrary stances on topics such as political, religious, or sexual inclinations, may wish—or even be compelled—to keep these aspects of their characters private. Yet, at various levels, these hidden traits probably affect almost everything they think, say, and do. For example, we can openly express empathy even while its roots deep within ourselves may remain unseen. An embarrassing or regrettable experience, which we'd prefer to keep secret, can nonetheless inspire us to be compassionate to others in similar predicaments.

Similarly, private data is an essential component of humem nature. Since a humem is a reflection and extension of its alpha's character, how could this be otherwise? Like people, humems should be able to conceal their private data while still being able to outwardly express the data's influences—the external value of the inner self. Again we see that it is misleading to view humems simply as a type of archive to which others may or may not have access. Humems are more like individuals whose characters, cultural surroundings, and the spirit of the times influence what they share and with whom. If humems are not able to ensure privacy, they will not be entrusted with personal or confidential knowledge. Consequently, they will not be in a position to perform many of the most compelling functions that we will desire of them. For one, they cannot help us remember that which they themselves do not know.

In current systems, private parts of our EPs have a precarious existence as discrete entities. They usually have no integral shields or filtering mechanisms that facilitate controlled external expression. Therefore, they are often either completely hidden or completely exposed, respectively making them either useless or detrimental. In reaction, many people, erring on the side of caution, try to constrain the creation of their private EP. In particular, they may desire systems that ensure the complete erasure of such material under certain circumstances—their death or loss of mental faculties, for example. Unfortunately, to prevent the formation of these aspects of their EP or to delete them later also eliminates a large part of the EP's core character.

Thus, in order to exist and flourish, one's private EP needs protections. Since there is no fixed boundary between the private and public EP, in effect, a large part of one's EP requires some level of shielding. But you cannot protect that which you are unable to identify. Accordingly, for this reason too, one's EP needs to be consolidated and recognized by its formation into a humem. Moreover, for these protections to be lasting—lifelong and beyond—they ultimately need to stand independently of the alpha-person. That is, the humem itself must be endowed with rights to enduring privacy. To attain these long-term goals, these provisions cannot be static. Instead, they need to be continually adapted to changing norms and environments.

A competent humem, possessing humem rights in an appropriate administrative environment, can facilitate the perpetuation of a private character in a manner similar to how an able person manages their own privacy of mind. While legislation is ultimately required to consolidate aspirations in this regard, a great deal can be achieved by technical means alone. These include techniques for the encapsulation of personal information and the control of the sharing of semi-private data with limited groups.[40] Such mechanisms are first valuable to the alpha-person for pragmatic lifelong uses, and thereafter can ensure the seamless continuance of such safeguards in their ancestral humem.

The humem administration will be responsible for procuring, applying, and updating the technology that will provide humems with these abilities. Furthermore, the humem-state will establish regulations within its jurisdiction to protect humems' rights to privacy and will also cooperate with other people and humem organizations toward similar goals. Consequently, the ability to guarantee a lasting private EP as the core of humem character and individuality will be one of the principal criteria for assessing the suitability of the proposed forms of humem governance. In addition to the more general aspirations for EP consolidation and perpetuity, privacy concerns will arguably be one of the strongest and most urgent motivators for many people to seek the establishment of a robust humem environment.

The Right to Freedom of Expression

By its very nature, humemity will be freer in expression than almost all current nation-states. Once an entity cannot be constrained within physical borders, it becomes very difficult to censor.

In direct contravention of the UDHR,[41] the current, primary EP communication highway, the Internet, is policed and significantly restricted by some nation-states. However, it appears that these states' ability to maintain such controls are gradually being eroded by the proliferation of new communication channels and other methods of circumvention. These new technologies facilitate a scope of personal freedom and

---

[40] The selective sharing of information between social network proto-humems, for example, is already well established.

[41] UDHR-Article 19: Everyone has the right to freedom of opinion and expression; this right includes freedom to hold opinions without interference and to seek, receive and impart information and ideas through any media and regardless of frontiers.

expression far greater than anything we have seen in the past. All indications imply that this trend will continue.

Thus, by and large, it seems that most humems will be born into environments that are sufficiently liberal for the freedom of mind and opinion that they desire. It appears that for the majority of humems, as for their alpha-people, any discord with their administrative bodies is more likely to focus on the contents of their pockets[42] than the contents of their minds.

The Right to Migration and the Choice of Jurisdiction

Over history, timely migrations have often shielded individuals from harm or have even been instrumental to the survival of large groups of people. In less critical instances, migration has allowed many to prosper in more favorable environments. The UDHR explicitly recognizes the importance of relocation and asylum in articles 13 and 14.[43]

Likewise, or perhaps even more, humems' ability to migrate with their assets to more favorable jurisdictions will be crucial for their longevity and ongoing prosperity. This is especially pertinent given that the longevity of an individual humem may, by far, exceed the typical *optimal life span* of a particular humem-state (the duration for which the state remains the optimal environment for humems' purposes or the period during which more favorable jurisdictions have not yet emerged).

As we discussed earlier, humems, at least in advanced implementations, will be mostly location-independent. So when we use the terms

---

[42] The "contents of humems' pockets" will be discussed in depth in chapter 8 on humem economy.

[43] UDHR-Article 13: (1) Everyone has the right to freedom of movement and residence within the borders of each state. (2) Everyone has the right to leave any country, including his own, and to return to his country.

UDHR-Article 14: (1) Everyone has the right to seek and to enjoy in other countries asylum from persecution. (2) This right may not be invoked in the case of prosecutions genuinely arising from non-political crimes or from acts contrary to the purposes and principles of the United Nations.

"jurisdiction" and "migrate," we need to consider carefully what we mean. Early on, humem jurisdictions will commonly be hosted within nation-states and will therefore be subservient to existing systems. As we will see later when we study the ideal forms of humem administration in more detail, an initial inclusion in a country does not necessarily preclude subsequent freedom from geographical constraints. In the humem context, jurisdiction will become more of an organizational, managerial, and legal construct than anything relating to geopolitical divisions. Broadly speaking, the jurisdiction "in" which a humem resides is the entity under whose authority the humem exists, and which is responsible for the humem's well-being.

To get a more solid feel for what this means, the following rough and limited-in-context analogy may be helpful. For the current purpose, let's compare a humem jurisdiction to a multinational bank and a humem "resident" to a personal bank account. (We can think of a personal bank account as an entity affiliated with a person, composed of data and assets, and possessing identity and a legal standing.) The bank as a whole is not really located in any one place. It is largely meaningless to ask where it keeps its assets or where its accounts are located, since these are almost completely virtual constructs. Granted, it may have many branches in which its employees work, and which function as its interface with the world, but it also has Web interfaces that are truly location-independent. A specific bank account is also mostly detached from any physical place. Although we say that the account is in the bank, there is nowhere one can go to actually "see" the account.

According to this analogy, different humem jurisdictions are akin to different banks, and the "emigration" of a humem from one jurisdiction to the other is comparable to the transfer of the assets in an account in one bank to an account in the other. The contents of an account—say balances in a particular currency or other securities—can be transferred mostly intact from one bank to the other. Typically, nothing material moves: electronic messaging (the transfer of data) and associated administrative actions are the primary dynamic processes.

While humem bodies will contain a much greater diversity than the contents of bank accounts, it seems that similar dynamics will be involved in humem migration. This analogy has additional relevance in that once humems have intrinsic value and own assets, their migration will likely also involve regulations relating to money transfer.

The Right to Marriage and Procreation

In chapter 4 we saw that, as emulations of people and their relationships, humems will also form connections that mirror those of their alphas. For example, the alpha-humems of a married couple will typically emulate the relationship between the people. Like a traditional marriage, the humem marriage will typically involve very tangible components such as shared resources, common ownerships, and mutual obligations. Just as humems can extend and enhance the existence of individuals, they can also enrich an association such as a marriage. Therefore, the right of two humems to institute and maintain such an association may also be beneficial to the people's relationship.

Furthermore, since individual humems also have intrinsic value, a marriage relationship between two humems may also have an innate value. In certain settings, it may be desirable for such an affiliation to be founded, or to continue, independently of the life-status of the people. For instance, if one or both of the alpha-people are deceased, their humems' existing relationship may still endure indefinitely. It is even conceivable that a new marriage or other relationship may be established with an ancestral humem (one that no longer has a living alpha-person).

Marriage is just one example of how humems will emulate people's relationships. Numerous other humem associations will also be possible. The creation of a child humem, formed by the merging of contributions from parent humems, may also be a desirable outcome. Such a humem may, or may not, be the alpha of a child person. The right to create such a child humem, with provisions for the transfer of resources from the parents for its sustenance, would also need to be ensured and regulated. Such a relationship involves many of the concerns relating to "real" parent-child associations. For one, laws in most jurisdictions determine the duration of the parents' responsibilities and how a child can be bequeathed assets. Similar considerations will prevail in the humem domain. We will discuss some examples of humem marriage and procreation in chapter 9.

The Right to Ownership of Property and Economic Independence

The rights to generate income and to own property are necessary for economic viability and self-sufficiency. Moreover, such economic independence is the precursor of further forms of individual autonomy.

Currently, proto-humems themselves are considered property. Like slaves, their inherent value and any income that they generate belong to their human or corporate masters. As such, their futures are shackled to those of specific people or companies. Any weaknesses in their owners' economic positions also undermine their prospects for continuation. When the corporations eventually dissolve or when their alpha-people inevitably die, they are left destitute and vulnerable. Their alphas' children or other kin may perhaps extend some charity, but after a generation or two this support will most likely diminish, and they will steadily decline into insignificance. The absence of proto-humem rights to economic autonomy is a major impediment to their long-term subsistence that must be rectified if they are to achieve lasting viability.

As we'll see in chapter 8, certain proto-humems already demonstrate economic capacities. We will explore some of the ways in which they may soon be able to sustain their own existences, and how humems may eventually diversify their means of income to mirror, and even supplement, people's paths to livelihood. Humems' ability to generate value, coupled with their rights to own and utilize the proceeds of such income, will play a vital role in their prospects for development and continuity.

The Right to Development and Social Security

In the first stages, most humems will be strongly dependent on human benefactors. Clearly, alpha-people and their families will be the primary sponsors of their related humems. However, for ancestral humems in particular, these human relatives will not always be able or willing to provide support. Therefore, to ensure their long-term survival, humems will require support systems and forms of insurance that are contained wholly within humemity. These can be regarded as social security rights within the humem-state. Let's touch on a few examples to get a sense of what this entails. We'll revisit this topic in more detail when we examine the humem-state and the humem economy.

Let's first consider how states ensure the welfare of their young. In modern societies the sustenance and schooling of children is usually seen as the dual responsibility of parents and state. Generally, the parents assume primary responsibility for the children's care and education, while the state provides the infrastructure that enables such development.

The state provides health services, schools and curricula, clean water, food safety regulations and their enforcement, and a myriad of other services that are essential for children's optimal growth. The state's acquisition of new technologies is also a vital process that ensures the survival and prosperity of its citizens. Modern states incessantly procure, develop, and standardize apparatuses for defense, health services, agriculture, and many other economic endeavors. The state also enacts laws for the protection of children and adults. In cases where parents or other relatives are unable to care for children, the state steps in and assumes primary responsibility for their welfare. Also, social security institutions function as safety nets for ailing individuals who have no other recourse.

Now let's translate this into humem terms, from the perspective of the humem childhood. In an appropriate humem environment, the alpha-people or other humem creators, in conjunction with the humem administration, oversee the development of the early humems. The people provide for the immature humems' individual needs, such as the resources required for their establishment and upkeep. The humem-state is responsible for protecting humems' health and for facilitating their growth by creating and maintaining a suitable environment. The state's services include the procurement and application of technology that comprises the humems' new minds and memories: in current terms—hardware, software, and the processes for their implementation and maintenance. It is not practical or economically viable for individuals to evaluate such technologies alone. However, as with existing people-state institutions, dedicated humem-state bodies can routinely perform such assessments proficiently and in a cost effective way. Competent standards bodies that leverage an economy of scale to continuously evaluate new humem-serving technologies are comparable, for example, to the US Food and Drug Administration (FDA) in their evaluation of new treatments. The humem-state also provides a system of social security for destitute humems for circumstances in which their alpha-people or other relatives are not able to care for them.

Nation-states establish militaries to protect the safety and interests of their citizens. Comparably, the humem-state adopts processes and technologies to safeguard the humem population and its infrastructure. In today's terms, these primarily comprise information technology security systems. This analogy to a military is especially pertinent as we observe conventional militaries expending progressively more effort toward developing various forms of cyber-security.

Importantly, the humem-state is not a closed system. For their development, humems will require ongoing interactions with people and with humems in other humem-states. Initially, at least, technologies and other humem-related services will mostly be acquired from outside the humem-state. To this end, the humem-state will also need to facilitate "international" relations—relations with institutions in nation-states and humem organizations in other humem-states. The former kind of transnational relations will be especially crucial during the early stages of humemity, when the humem-state will be one in which almost all its citizens have close relatives in nation-states (such as their alpha-people or alpha-people's families).

This brief glimpse of the future humem environment, and the services provided by the humem-state, provides a sense of what the rights to development and social security may mean in the humem context. In essence, they are very similar to those of people in advanced nation-states. Like most modern democratic nations, even though not all humems will be equally affluent, they should all have the right of access to the basic services and protections provided by the state.

There will, however, be some aspects of these humem rights without precedents in existing frameworks. One example is humems' eligibility for certain rights to be independent of their ages or the life status of their alphas. For entities with open-ended life spans this may involve different considerations from those of people. In particular, the rights of ancestral humems (those without a living alpha-person) to an assured and continual existence will be a central theme in the humem world.

Humem Ethics and Obligations

With citizenship come not only rights but also obligations. As an integral part of how they will emulate and enrich human behavior, humems will largely adopt the ethics and customs of their alphas.

The basic framework for humems' ethical conduct, like that of people's, will be shaped by the prevailing cultural norms of the time. At a formal level, many of the guidelines for humem behavior will be described in the regulations and laws of the humem-state. These, in turn, will be strongly influenced by those of the associated people-state, which is typically the state in which the majority of the humems' alphas reside.

Early manifestations of the convergence of the alpha-humem's and alpha-person's respective characters are already evident. Observing the current social network proto-humems, we see that in most respects they reflect their alphas' personalities. They assume their alphas' politics, religious views, cultural values, and even sense of humor. Online etiquette and accepted behavioral norms are now becoming established in social networking communities. In some settings, these behavioral dynamics are still reminiscent of those of the Wild West, sometimes with self- or locally-appointed law enforcers, or moderators. But in others, signs of maturity with increasingly accepted norms of conduct are emerging.

In their current implementations, proto-humems are often like puppets directed by their alpha-people from behind screens. Thus, their conduct directly reflects their alphas' ethics. However, as humems advance and gain their footing, they will behave more independently:

more like personal assistants and alter-egos, still closely influenced by their alpha-people, but with increasingly self-regulating and capable characters.

In childhood, the humem will have limited capabilities, and thus, less responsibility. The alpha-person or other custodians may perform certain obligations, such as the payment of fees to the humem-state. When humemity matures and individual humems reach adulthood, they will be required to more independently adhere to the cultural norms and to uphold the laws and ethics of the humem-state.

To ensure the continuation of humems' appeal to people, their morality will require a strong sense of devotion to their relatives. Humems will be expected to be *loyal*, especially to their alphas and other close kin. This humem attribute is crucial for their long-term prospects. To safeguard against humems becoming outlaws or pirates, humem-states, like people-states, will play a decisive role, both during the humem childhood and at later stages.

Social Networks and Emergent Discontent—Modern Deprived Humems

We have discussed examples of social networking applications in various settings and shown that they exhibit the early stages of humem behavior. These kinds of proto-humems are growing rapidly, gaining capabilities, and permeating ever deeper into people's private and professional lives. We can therefore reasonably ask whether they constitute a suitable and constructive first step toward the establishment of more complete and competent alpha-humems.

In our examination of the characteristics of these proto-humems and the needs for humem rights, we have already touched on a number of drawbacks to this approach. To further highlight these deficiencies, and to assess the general idea of a social network environment acting as the proto-state for these new humem-citizens, at this point I propose considering an alpha-person's metaphorical account of their early experiences within the social network. This symbolic narrative is intended to summarize and consolidate many of the concepts we've explored thus far in preparation for the next chapter's analysis of the appropriate type of humem governance.

Humem Playgrounds and the Not-So-Free Ride

Technology has established a digital primordial soup in which the synthesis of our proto-humems has been taking place. These fledgling beings have been erratically emerging in ad-hoc forms within our modern machines. They have been developing in the stifling confines of our personal computers, and in semi-isolation in compartments such as personal webpages and blogs.

Born from people, they are, at heart, social creatures. However, autonomous interaction between these rudimentary proto-humems has been awkward or impossible due to their widely divergent appearances, dialects, and capabilities. We, their guardians, have needed to facilitate all their communications. As our proto-humems are becoming smarter, more endearing, and closer to us, we almost cannot imagine how it was without them. But these nascent beings, each different from the others, have been growing at an astounding pace, and we have been unsure of how to take care of them.

Recently, though, state-of-the-art social network playgrounds catering to a generic form of proto-humems have been created. Following a few partially successful attempts, a really popular one has emerged. It offers an appealing solution for our young progeny: little exoskeletons into which we can insert them. Their now homogeneous exteriors have standardized communication and sensing apparatuses, which allow our proto-humems to find each other, make new friends, and interact, while adhering to the playground's conventions.

These engaging beings, ensconced in their protective and reassuring exoskeletons, are participating freely and developing rich social lives. We, their guardians, with them in hand, are having a wonderful time too—meeting old friends and making new acquaintances in this lively and colorful environment. Here, many of us look our best and feel our most confident. Our social interactions are better than ever. In fact, we are often unable to imagine how we could manage without this playground—it is free and far better than the alternatives.

After the initial euphoria, some of us—still a minority—have begun having concerns relating to our humems' welfare. When we first visited the playground, our humems were still quite small and underdeveloped, so we did not worry too much about the future. But now they are becoming dearer to us, almost like children, and therefore we are paying closer attention and thinking a little further ahead. We have begun to notice some troublesome signs.

In our first excitement, following our friends, few of us really read the release form, or agreement, that we signed at the playground's entrance. But some of us did. It stated that, although the playground management would most likely never separate us, once our humems entered the playground they were no longer completely ours; actually, in a sense, they formally belong to the playground. The agreement stipulated that as long as we behaved sensibly, we could probably always interact with them in the playground. Yet in this regard, none of us are quite sure what "sensibly" and "probably" and "always" mean.

Also, we have noticed certain limitations. Our humems are not allowed to know or think all the things that we do. They have to be polite and well behaved; even internally, they are not permitted inappropriate knowledge. Thus, they are not as complete or as similar to us as we would want them to be.

The agreement also mentioned that for various transgressions, or just at the sole discretion of the management, we could be banned from entering the playground ever again. Most of us do not personally know anyone who has had a major problem with these rules, but this still causes us worry. Our humems are becoming very habituated to this particular place; already, although they have not been here for very long, they have become molded to the inside of the exoskeletons that the playground provided. If we ever had to leave with them, they might have great difficulty migrating and could even be damaged. And how would they keep in contact with all their new friends?

At some time in our lives, most of us have learned that there is no such thing as a "free ride." So while it has been upsetting, it has not been a complete surprise to notice several other disturbing developments. The playground is dotted with vendors selling products for both the humems and for us, their guardians.

Since we have not really been strongly coerced into buying anything, we are not overly bothered by their presence. However, we have been perturbed to discover that the vendors seem to know us personally. Somehow, they are aware of the content of our interactions with other guardians and humems, and our private communications with our own humems. It seems that the playground's administrators are able to peer into our humems' minds and memories; thus, we are hesitant to entrust our humems with very personal information such as medical, financial, or intimate family matters. This is unfortunate because the closer the humems become to us, and the smarter they get, the more we'd like to share with them.

Even though we and our humems entered the social network playground voluntarily, we feel a bit trapped. The playground is becoming our humems' home: they spend all their time there, all their friends are there, and they do not know how to exist anywhere else. Since we have formally relinquished the ownership of our humems, they are, in a way, like orphans without lawful parents. In fact, the playground provides them with a kind of "room and board" in exchange for getting to know us better, in order to sell us things. As a commercial enterprise, its primary motive is to maximize profit for its investors.

The playground manager seems like a well-meaning person who cares about his patrons' welfare. Nevertheless, he and his advisers determine our humems' futures; we, as patrons, have little say in the outcome. Despite the manager's good intentions, he may be forced into policies that are not in our humems' best interest. Furthermore, he will not hold his position indefinitely and may be superseded by someone less amiable to our humems' welfare. Since the playground is becoming the humems' only world, with its own society, conventions, and administration, it increasingly seems like a kind of state. When viewed in this way, this state's form of governance is comparable to a monarchy or dictatorship.

As our humems develop, they are becoming more similar to us. And, the more the humems resemble us, the more we want them to have comparable benefits and opportunities to ours. We are also beginning to realize that they can continue to grow and persist and interact with our family and friends long after our bodily decline—we do not want them to diminish or expire just because we will.

But how can the social network playground—a commercial body— ensure the humems' long-term continuity? It seems clear that our humems' habitat should not be a commercial entity at all but rather an environment resembling ours, where our humems can have rights that are similar to our own but that are adapted to humems' special needs. The humem's environment should have an administration that mirrors that of our nation-state, with internal checks and balances, and institutions that are devoted solely to humem welfare—a dedicated humem-government within a humem-state. The social network playground does not meet these criteria and has no plans to do so. A business cannot expand into an idealistic, adaptable, and long-lasting humem administrative body—these two kinds of institutions are based on fundamentally irreconcilable principles.

Clearly, this vision of a dedicated humem-state has a cost, but it also has immense value: humems will sustain the humem-state in the same way that our own economic activities and taxation fuel our nation-states' functions. And while our alpha-humems are becoming increasingly precious to us, they are likewise valuable to numerous commercial entities, such as the playground, whose huge and rising revenues are derived primarily from our humems' expressions. A dedicated humem-state will redirect much of this emerging value to directly benefit the humems and their alphas, and thereby empower the alpha-pairs to determine their own destinies.

For now, we and our humems will continue to participate in the playground since abstention would severely disrupt our social lives. Nevertheless, the playground should be regarded as such, and not as our *humems' permanent abode. Our humems should be able to visit or migrate to other environments and to develop their full potential, unconstrained by the limitations of content and expression, and an uncertain future imposed by the playground. The alpha-pairs should be able to own the proceeds of the value that they generate and to direct this affluence toward their self-defined purposes. To this end, we need to establish a primary humem habitat—the humem-state—that will serve our humems' long-term needs, ensure their rights and security, and recognize them for what they are and what they can become.*

CHAPTER 7

GOVERNANCE AND STATE

The previous chapters described humems' emergence, characteristics, and initial modes of interaction with people, and the justification for endowing them with rights and citizenship. As I previously asserted, citizenship implies governance. In this chapter, after further evaluating the necessity of a dedicated and independent humem administration, we will consider which types of humem-government and humem-state are most conducive to humem needs and aspirations.

The Practical Need for Governance

Nations in today's interconnected world are unable to function properly without stable, well-functioning governments. Yet, in modern societies, having a government is not a goal or an ideal unto itself: our ideologies—religious, moral, or political—do not usually dictate, "And thou shall create a government which thou shall cherish and love...," and so on. Rather, most of us view a government as the best existing means for ensuring our safety and for advancing our welfare and aspirations. If alternative structures were to emerge that proved more advantageous to the needs of citizens and societies, they would be duly adopted. Good governments, and especially good democracies, are designed to facilitate progress and smooth migration toward improved systems.

In this sense, what holds true for people also holds true for humems: the establishment of a government-like body is not an intrinsic objective but rather the most practical way to organize, protect, and develop humems—as individuals and societies. In fact, it is the similarity of humem-citizens to people-citizens, and humems' multifaceted interactions with people and their institutions, that prescribe the need for a type of humem governance comparable to that of people. Initially, humems will be deeply dependent on the infrastructure and services of developed states. Thus, humem-citizens will require governance that can both coexist with traditional governments and accommodate humems' special requirements. Just as people need well-functioning nation-state governments, humems need an administrative environment that is designed to accommodate continual adaptation and improvement.

Looking around today, and peering as far back into history as we are able to, we can see that modern democracies have opened the door to more personal freedoms than were ever available to the majority of people in the past. Of course, there are significant differences in the way people perceive personal freedoms and rights. Still, there is a fairly broad current consensus on this—as expressed, for example, in the Universal Declaration of Human Rights (UDHR). In fact, *most* individuals in modern democracies concur with *most* of the UDHR clauses.

Some people consider an absence of governance as being the most conducive to personal liberty. However, in the real world, there are few pristine jungles or deserts where individuals can supposedly gain such freedoms. Wherever such places remain, the inhabitants are vulnerable and their futures uncertain. By contrast, deliverance from the persistent fear of existential danger is a fundamental freedom that is best provided by good governments in stable states.

Similar considerations pertaining to freedom and safety are also relevant for humems and are especially decisive during their vulnerable childhood. The protections and services afforded by good humem-states will, by far, offset the restrictions that are commonly imposed by governing bodies. Moreover, limitations to humem autonomy, where they exist, will frequently be transitory and will gradually dissolve as humems develop maturity. Again, this view is consistent with the previously introduced concept of the childhood of humemity, which implies a safeguarded environment combined with temporary constraints. Therefore, based on the presumption that humems do need governance within a dedicated environment, let's now consider its ideal form.

Forms of Governance

In this section we will examine a number of apparent candidates for humem governance. For each, we will appraise both its theoretical suitability for humem administration and the feasibility of its establishment.

Commercial Citizens—Corporations Administering Humem-Citizens

To begin, I will make a simple case for exclusion. Corporations, which already play a dominant role in the proto-humem environment, should be categorically eliminated as serious contenders for humem governance. You will remember in an earlier discussion (Social Networks and Emergent Discontent—Modern Deprived Humems) that we examined a number of reasons why this is the case. In essence, the conflicts of interest are too numerous and commercial organizations are typically too short-lived. Even the few corporations that have endured for a long time, such as a century or more, have generally failed to maintain a consistency of purpose for the duration of their existence.

The argument against corporate governance should be entirely intuitive to educated citizens of modern democracies, who would not seriously consider having a nation-state owned by, or under the management of, a corporation. For precisely the same reasons, such a solution is incompatible with humems' higher aspirations and their long-term welfare. As is true for successful nation-states, corporations will certainly play a central role in providing services, driving innovation, and building infrastructure for humems. However, the overall management, ultimate responsibility, and any form of ownership of humems cannot be placed in corporate hands.

Piggyback Citizens—Humem-Citizenship Appended to Person-Citizenship

If we rule out corporate governance, where do we turn? Before envisioning an entirely new system, it makes sense to consider whether existing mechanisms could be gradually adjusted to facilitate humem needs. In particular, one could ask if the legal status of people—as recognized by the people-state—could be amended and expanded to include their alpha-humems. To assess this approach, let's first examine how present states relate to existing proto-humems. Then we can try to extend this understanding to predict how nation-states might interact with humems in the future.

In previous chapters, we explored the earliest form of proto-humems: the consolidations of our EPs—the models of us—in others' brains. How do states, then, treat the content of their citizens' minds? Well, we know that our thoughts have protections in most developed countries. Some, like the USA, by virtue of the Fifth Amendment, explicitly protect the privacy of the mind in certain circumstances, such as one's right to remain silent in the face of criminal accusation. Also, in accordance with the Universal Declaration of Human Rights, most jurisdictions prohibit attempts—by means of torture, for example—to forcefully access the contents of one's mind. Other less extreme methods of achieving similar results, using techniques or tools such as polygraph tests, are controversial but sometimes permitted.

It is clear that most of us abhor the prospect of others having direct visibility into our minds, and yet recent advances in neuroscience indicate that this will be a reality in the coming years. For example, brain fingerprinting, a forensic technique that uses electroencephalography (EEG) to determine whether specific information is stored in a subject's brain, is already making its first appearances in US courts.[44] Many existing citizen-privacy practices remain contentious, while, driven by technology, complex new problems are continually emerging.

---

[44] Brain fingerprinting, using electroencephalography (EEG), was first ruled admissible in an Iowa (US) court in the early 2000s.

It is also apparent that an increasing portion of our memories will reside outside our physical minds. They will either be part of our fragmented EP, or as proposed in this book, incorporated in our alpha-humems. These memories, though materially apart from us, will be as essential as those held within our minds. Due to their immense size, exquisite detail, and longevity, they may be even more precious in some matters than our internal memories, and arguably in greater need of protection.

For example, we may legitimately claim that we don't remember some elements of an experience. Our alpha-humems, however, will recall certain aspects of past events in minute detail. They will be in a position as well, to substantiate their assertions in ways that we cannot. For example, they will typically have the ability to present the original data that they store in their memories. Thus, most of us will want our humems to enjoy at least the same level of *privacy of mind* as we ourselves do. The alternatives, either opening up our intimate knowledge to others or severely curtailing the development of our humems, will be unacceptable.

Yet, at least for the foreseeable future, it is clearly unrealistic to expect existing nation-states to extend similar protections to abiotic-humems—artificial entities in their eyes—to those they currently confer on the contents of people's minds. In particular, if states consider humems to be nothing more than addendums to their alpha-people, regulated by means of modifications to the current legal status of citizens, it is difficult to imagine how humems' legal standings could be viewed as anything other than property.

Consequently, this would impart humems with the inherent vulnerabilities of possessions, which can, for example, be seized or forfeited in situations where debts have been incurred. In civilized societies it is generally accepted that a person's body and mind, or their offspring and spouse, cannot be seized in lieu of debt or as a form of punishment. But property and other assets are more vulnerable. As alpha-humems mature, their emotional value to us and their potential monetary value to others will grow. We will perceive them in the way we perceive our own minds or our close relatives. With that, it will be inconceivable to us that they be susceptible to loss like property.

For most people, alpha-humems will become useful and necessary lifelong consorts. For others, however, humems' prospects for permanence will constitute their primary appeal. If humems are legally regarded as possessions, their potential longevity will be severely reduced; in addition to the humem's vulnerabilities during its alpha's lifetime, an entirely new set of dangers arises with the alpha's death: Complicated estate laws come in to play; property is coveted by beneficiaries and creditors, and may be subject to taxation; lawyers become involved. None of these entities necessarily have the humems' best interests in mind.

Prudent parents do not wish, after their death, to leave their children destitute, helpless, and devoid of rights, simply hoping that the provisions of their wills and the good services of their executors will set things right. Individuals and organizations require an extensive period to develop self-sufficiency and resiliency. Similarly, it is unrealistic to anticipate a situation in which the partners of the alpha-pair exist together, with a thriving person and a dependent humem, until the person's death, when somehow a mechanism kicks in to emancipate the humem. A viable humem should have most, if not all, of its abilities and resources in place long before it needs to "stand on its own feet."

But even if it were reliably possible to establish an autonomous ancestral humem after its alpha-person's death, an earlier and more disturbing risk would still remain. In the modern world, sudden death is becoming an increasingly rare occurrence. It is far more common for people to suffer physical, cognitive, and economic infirmities as they near the end of their lives. In these circumstances of slow demise, the person's remaining dependents are at risk—including a dependent humem.

Many conflicts of interest could arise. For example, perhaps without the alpha-person's consent, resources destined for the humem's future sustenance might instead be expropriated for medical expenses. As one approaches the end of one's bodily life, this could culminate in "selling one's soul" in exchange for a few days' delay of the inevitable.

Instead, we can turn the dependence scenario around and look at it from an opposing perspective. Not only do resiliency and independence increase the alpha-humem's prospects for posterity and success, but these same characteristics also empower it to assist the person in their time of need.

A modern person's decline can often last for many years. This should be a period in which an autonomous and able humem, like an independent and capable child, assists and serves (and, in a sense, repays) its alpha-person. In the person's final years, the humem can be especially helpful in augmenting their mind and memory, and in facilitating their interactions with the world. Earlier in life, the person might have been the dominant partner relating to matters of character and core decisions. But later, particularly if the person becomes mentally frail, the humem may become the more competent one—supporting its alpha with a knowledge and sensitivity gained from an intimate, lifelong partnership.

With these considerations in mind, we can conclude that, in current settings, it is neither feasible nor desirable to append the alpha-humem's formal standing—its citizenship—to that of its alpha-person. Doing so would shackle them together in a way that would inherently weaken and endanger the humem, without benefiting the person. An adjunct, or piggyback, citizenship denies the humem its individual character and capabilities, and inextricably entangles the humem's prospects for the future with the person's mortality.

Segregated Citizens—Separate Humem-Citizens in People-States

There may be another way to integrate humem-citizens into people-states. Perhaps humem-citizenships could be embedded within the frameworks of existing nation-states while remaining *formally distinct* from the alpha-people's citizenships.

Before looking at this idea more closely, we should note that, due to the nature of current governments, most of the objections described in the previous section apply here as well. The recognition of humem-citizens as separate entities within current states poses a number of additional problems. The reality is that many people-government administrative principles are fundamentally irreconcilable with humem requirements, and it is unlikely that this will change anytime soon.

For instance, people-governments are typically driven by short-term goals. They are often focused on the upcoming elections, or if they are particularly savvy, they may plan a decade or two ahead. As a result, their policies are mostly dictated by people's immediate concerns and clearly perceived needs. Humems, however, have different require-ments and are especially sensitive to more distant considerations, such as their economic sustainability over extended periods. Unlike people, humems, by their very nature, will tend to defer transient gratification in exchange for improved long-term prospects.

Also, location-centricity, where it still applies to humems, is steadily diminishing in significance. Humems will gradually become almost completely location-independent—a condition that is essential for their long-term viability. In practice, existing governmental institutions will have difficulty accommodating a new kind of non-localized entity, while continuing to serve location-centric people.

Another primary obstacle to the "inclusion in the current nation-state" approach is the startup catch. In practical terms, it will be very difficult to persuade governments to effect humem-friendly legislation before humems are already well established and politically significant. But, to become established and influential, humems first require the appropriate institutions and legislative environment.

Especially early on, the humem condition will be crucially dependent on interactions with people-states. Therefore, a dedicated administra-tion—one that is dissociated from the people-government—will be a much better advocate and negotiator for the benefit of humems.

In many unrelated systems, such as aged buildings, damaged cars, or rigid political doctrines, a stage is reached when it becomes much more efficient to build a completely new structure than to fix the old one. The same appears to be true for the current people-governing agencies: it will probably be far more effective to create new humem-governing organizations than to adapt the old ones to the humem purpose.

Even if somehow separate humem-citizenships in people-states were established, existing governments would always remain biased in favor of the people-citizens they are appointed to represent. Humems would remain politically segregated and suffer the discrimination of second-class citizens. Due to the intimacy between the members of the alpha-pairs and the participants in other modes of close people-humem relationships, forms of segregation under a unified government would be detrimental to all. Over time, as humems become dearer to us, such disparities will become at least as repugnant as other historical episodes of discrimination.

In a similar vein, future humems may be able to democratically influence their governance. But could this work under a single administration? Humems and people will often have different concerns, especially due to their sensitivity to very different time frames. Therefore, it is hard to imagine how they could participate as peers under a single democracy.

Yet, even if, in one way or another, forms of equality or divisions of influence were achieved and accepted, a fundamental disqualifier would still remain: the *demographic problem.* Humem populations will grow continuously. Furthermore, they will be composed of an increasingly large proportion of ancestral humems. But, in parallel, people populations will stay more or less constant—or at any rate, they will grow more slowly than the humem populations. Consequently, humems will become the majority. Hence, democracy—at least from the people's perspective—would not work.

In this section, and the previous one, we have examined only a small sampling of the many complications attendant to establishing humem-citizens under their alphas' people-governments. Wherever we look, we see conflicts of interest and divergences in the requirements of people and humems. Nothing seems to work or make sense. Accordingly, we can conclude that, at least in the foreseeable future, people and humems need separate governing bodies.

Offshore Citizens—Humem-Citizens in Foreign People-States

We can easily imagine how offshore[45] jurisdictions could provide facilities for affluent humems. Such dominions are often very agile in offering innovative financial solutions and rapidly adapting their legislations for the sake of lucrative commercial interests. While such agility may be very attractive during the early days of humem establishment, these same characteristics pose a long-term risk to humem welfare.

First, most of the objections to a corporation fulfilling the humem administration function are also valid here. Whenever the interest is primarily of a commercial nature, the same types of conflicts of interest can arise. Moreover, the same plasticity that allows an offshore type of government to rapidly adapt a system to humem needs, because it does not have to answer to complex and conservative mechanisms of oversight, could just as swiftly be employed to enact undesirable policies under changing conditions or duress.

Relative to the hosting offshore jurisdiction, the humems' people-relations (e.g., their alpha-people) would be foreign citizens with little influence in counteracting such detrimental actions. The welfare of humems originating in other countries would most likely be of secondary consequence in any conflicts of interest between the humems' well-being and the interests of the offshore jurisdiction or its citizenry.

Clearly, most of the concerns presented in the previous sections regarding the administration of humems by the alpha-people's governments also apply here: humems are simply not the mainstay of traditional states. Therefore, it is apparent that people-governments, irrespective of their jurisdictions, are not suitable humem custodians.

---

[45] Here I use the word *offshore* in the context commonly associated with banking, i.e., in a foreign jurisdiction relative to the one in which the entity under discussion resides.

Needy Citizens—Nonprofit Organizations Administering Humem-Citizens

Taking the previous considerations into account, I suggest that humem administrations—like those of people—need to be founded on idealistic motives. This brings to mind nonprofit organizations, which typically excel at educational and charitable functions. And yet, while they often have idealistic purposes, nonprofits are usually limited in scope. Characteristically, they do not contain the mechanisms for scalability, adaptation, and self-correction that are required of good governments. Furthermore, they are not designed to develop into organizations dedicated to individual citizens with rights and rapidly evolving needs, or ones that encompass and administer versatile and self-regulating economies.

With good reason, nation-states are not equated with nonprofit organizations: nation-states and their governments have much broader and long-lasting mandates. Among their most fundamental functions are the protection of their citizens and the enabling of economic relations with other states. In no lesser way, humem administrations will assume similar responsibilities. Thus, to best serve the interests of their humem-citizens, humem-states will ultimately require the status of peer, rather than subordinate, to other states.

In the early days of humemity, some people may view humems, especially ancestral humems, as incapable objects or needy citizens that depend on the support of a charitable benefactor—the humem administration. Consequently, a charitable, nonprofit organization may seem suitable for this role. But as humems gain capabilities and independence, and become viable economic entities, the limitations of a nonprofit organization and the need for a more capable state-like system will become ever more apparent.

I should point out, however, that when we deliberate the initial mode of humem governance, we will find that nonprofit organizations may play an important tactical role in the establishment of the proto-humem-state. Nevertheless, this should be viewed solely as a pragmatic interim function and not as the form of the long-term solution.

The Natural Humem-State

After presenting several potential frameworks for humem governance, though in a cursory way, and arguing why each possesses fundamental and disqualifying flaws, I will now propose a suitable and feasible model for the humem-state. I will begin by describing the vision of the humem-citizen in the advanced, or what we can call the *ideal,* humem-state. Then, at more length, I will present a practical, interim solution—the first step toward this ideal state.

My intention in this discussion is to demonstrate that this ideal state is ultimately attainable—no insurmountable obstacles preclude its establishment. For it to come into being, there is no need for future inventions or fundamental changes in world affairs. Even as I describe the nature of this state at a conceptual level, I will map these abstractions onto tangible constructs using examples from existing institutions.

That said, I acknowledge that, due to an assortment of quite ordinary reasons, many of which are common to nation-states, this ideal state is probably not *immediately* achievable in its complete form. State building requires substantial time, effort, and resources. Historically, nascent states required extended periods to develop and prepare for independence. In the case of the humem-state, many organizational mechanisms need to be built from scratch, existing people-governments may contest the transfer of resources from their jurisdictions, and so on.

Even beyond that, there is another essential reason—perhaps the primary one—why the ideal state cannot or should not be imminently established: humems, because of their immaturity, are not yet able to autonomously populate such a state.

Moreover, for us people it is probably just as well. Even if humems and the humem-state were somehow soon ready to fulfill their full-blown destiny, we, their human relatives, would need time to adjust to their newfound independence. Cultural mind-sets need time to evolve. Thus, it is congruent with humems' present capabilities, and with our nature, that the humem-state first be established in an interim form much "closer to home."

Accordingly, following a fairly brief discussion of the ideal humem-state, we will proceed to lay out the structure of the practical and imminently achievable solution. A caution, though: while the initial system is a substantial undertaking that may actually suffice and endure for quite some time, we must constantly keep in mind the long-term vision of the ideal state. Our ultimate destination, however distant today, must already determine the direction of our first steps.

The Vision—an Independent and Locationless Humem-State

We have already acknowledged that government and state structures are not ends unto themselves. They are simply the most effective known systems for organizing present societies. The same is true for the proposed humem-state. Thus, in the following discussion, when I discuss the "ideal" humem-state, I refer not to an ideal state in any theoretical sense but rather to the optimal system for humem administration in the real world at any given time.

In many respects, the needs of humems and people are similar and interconnected, and therefore the ideal state for humems will have commonalities with well-functioning people-states. And yet, we know that there is no standard state or government that is optimal for all peoples. The distinctive nature and culture of the citizens determine the system that is best suited to their needs. Similarly, the origin of humems, which would typically correspond to the origin of their alphas, will strongly influence the conception of the ideal humem-state. For simplicity, we will assume in our analysis that the state in which alpha-people reside is a modern democracy. Other humem-states will be established based on the cultural givens of their corresponding people-states. With this understood, let us proceed.

Key Characteristics of the Ideal State

Notwithstanding the expected differences due to cultures and eras, several fundamental features will characterize the ideal system for humem administration. The humem-state should foster the essential humem rights of its citizens, be autonomous and self-dedicated, be sustainable over unlimited periods, and be location-independent. We discussed humem rights in some depth in the previous chapter. Let's now examine these additional state attributes.

- **Autonomy and Self-Dedication**

The humem-state's raison d'etre is first and foremost to benefit its humem-citizens. It should be primarily dedicated to this intrinsic purpose and to no other. It should be autonomous—neither subservient to the needs of other states, nor reliant on any benefactors external to itself.

Yet, if we pause and consider the humem-state as originating from people's efforts, a number of quandaries may arise. Naively, some people may consider a utilitarian motive as the main driver for the creation of humems. And if so, why should the humem-state not be subservient to a people-state? Certainly, the humem-state may produce many benefits for external entities such as nation-states, persons, and corporations. However, if the humem-state's utility to others is the primary justification or motivation for its existence, it will be prone to fail over the long term. This concept is entirely clear if we compare the humem-state to people-states.

In our earlier discussions about the EP and the appearance of proto-humems, we observed a number of phenomena that suggest that the EP can attain degrees of self-volition; it often appears to develop in itself, for itself, as do most living things. Also, we saw that in many respects, proto-humems are extensions of personalities, which in some forms can persevere independently of their originating people. Because they possess these characteristics, humems cannot be regarded as mere tools. Instead, they require the recognition of their inherent value. This mind-set needs to be ingrained in the humem governing institutions. It is the humems, above all, which they serve.

To better understand what we mean by an organization that is intrinsically dedicated to humem welfare, it may be helpful to look at some analogous, though simpler, examples. Let's compare the mission of an organization that provides guide dogs to blind people to that of an organization devoted to protecting gorillas in their natural environment. Presumably, the former values guide dogs (in this particular function) exclusively for their usefulness to people. If, miraculously, vision impairment in all its forms were cured, guide dogs as such would have no further use. The organization would cease to exist because guide dogs themselves were never its ultimate goal. A greater good—the welfare of blind people—was its primary purpose.

The institution for the protection of gorillas, however, is fundamentally different in nature. Assuming that it has no hidden agenda, it deems the ongoing existence of gorillas as having an intrinsic worth—one that, in essence, does not need justification in terms of its utilitarian value to people. The institution regards gorillas as having a right to be, and as long as they exist, the organization's primary mission remains unchanged. A similar worldview relating to the intrinsic value of humem existence is ultimately what is required in an enduring humem administration.

Still, the best analogy for the humem-state remains that of an independent people-state or society. To highlight this "intrinsic value of existence" principle let's consider the example of a reclusive state or culture, or an isolated tribe in a remote jungle, which has no discernable practical value to the external world. (For the sake of argument we'll say the culture is so reclusive as to even preclude anthropological study.) Nonetheless, by modern norms at least, there is no requirement for such an entity to be useful to other people or states as a justification, or prerequisite, for its very existence. Once it has come into being, it is considered to possess an unalienable right to persist. While humem-states can be shown to have enormous potential value for people, a similar intrinsic right to exist is necessary for their long-term viability. As another, somewhat extreme, example, humem this time, we can think of a future humem-state in which all its citizens are ancestral humems without close living human relatives. For this state, we could imagine circumstances in which there are no external entities with any particular interest in its continuation. Consequently, if its reason for being was measured primarily by its utility to others, and if it was dependent on these other parties, it would likely be prone to dissolve over time.

Granted, as we saw with the organization for the protection of gorillas, an administrative body can have an intrinsic purpose without being self-sufficient. But, as long as it is dependent on external support, it remains vulnerable to exploitation and neglect. Its sponsors' priorities, motivations, and abilities to sustain it may change. Its human supporters will eventually pass on, and may or may not be replaced by others with similar sentiments. Thus, like a good people-state, the humemstate must be self-governing, intrinsically self-interested—dedicated to its own citizens—and economically self-standing.

- **Sustainability**

There are many factors that influence the sustainability of the humemstate. Its autonomy (discussed above) is one crucial component, and its related economy (to be examined in the next chapter) is another. In this regard too, the humem-state should mirror good people-states and governments: the "why" should be stable and enduring, and the "how" should be nimble and adaptable. That is to say, its underlying purpose should be entrenched and stabilized through lasting directives, such as constitutions, while its operational methodologies should be designed for more rapid change. Its legislative and executive arms should be separated. It should have checks and balances for internal monitoring. There should be no single point of failure, and there should be ingrained mechanisms for the rectification of faults. Processes for development should include the routine replacement of personnel and operational mechanisms.

These elements of sustainability are similar to those of well-functioning nation-states. However, due to humems' potential longevity on the one hand, and their strong dependence on rapidly changing technology on the other, the consequence of each of these radically different timescales—consistency of purpose and variation of method—may be more extreme for humems than for people.

- **Location-Independence**

Although humems need states for many of the same reasons that people do, one thing that the humems clearly don't need is land. As we saw in chapter 5, humems are not geographically localized in any substantial way. The tangible parts that compose their bodies can be globally dispersed—often and preferably with replications of their parts. Moreover, humems' components can frequently change locations without detracting from their basic characters. Anchorage to a place and confinement within physical borders detract from humems' abilities and endanger their longevity. As such, we can say that humems are *locationless*[46] entities.

In a similar way, the ideal humem-state should also be locationless. Its infrastructure and government will clearly be based on tangible entities that exist somewhere—including people administrators, machines, and finances. Nevertheless, these components can be physically dispersed without fundamentally affecting the state's character. For the foreseeable future, humem-states will employ people, but it does not matter much where they live because their humem-state functions should be decoupled from their people-state citizenships.

That said, when we discuss the interim organization, we will see that location will most likely play a substantial role during the initial stages of state formation. For example, the location of humem monetary assets would have legal and tax implications, and so on. Over time, however, the significance of place will gradually diminish.

Feasibility of the Ideal State—a Case Study

So, in essence, aside from their locationless natures, we can envision the fundamentals of the humem-state and government mirroring those of good people-states and governments.[47] Nonetheless, the implementation of this humem system may be markedly different from those of existing states.

For example, when we think of existing governments, we tend to imagine huge bureaucracies. Yet, except for the legislative and senior administrative functions, most humem-government roles could be

---

[46] I introduce the adjective *locationless* to mean, "lacking affinity to geographical location." Superficially, this may seem similar to the term "landless." However, being landless is often understood as the condition of not owning or being affiliated with land but still being situated in a specific locale. By contrast, the term "locationless" implies that the described entity may never be localized, or located at a particular place, at any given time.

[47] *Good* here broadly means that, for significant durations, most of the citizens are generally satisfied with their government, or at least, they consider it superior than those of other states.

delegated to third parties—the equivalent of government contractors. As we have already seen, nation-state functions such as development, defense, health, and others, all will have humem counterparts. These can be assigned to commercial providers under the oversight of the humem-state administration, thereby keeping the core organization modest in size.

Are there any insurmountable obstacles to the creation of such a humem-state? Since the humem-state does not require land, the historically most contentious issue for the establishment of new states does not exist. Assuming that it does not adversely affect others, would anyone or any other state object to its creation? Is such a purpose-dedicated, locationless state, with close ties to many citizens in conventional people-states, really feasible? It turns out that there is at least one prominent example of such a state-like entity in the world today—one that has existed for centuries and seems poised to thrive for many more. Let's see what we can learn from the Catholic Church's Holy See, or the Vatican.

Before we do, though, let me first make it absolutely clear that I have no intention of equating the ultimate purpose of a humem-state to that of the Holy See. Likewise, it is not my aim to compare the types of governance of these organizations. However, there are a number of noteworthy similarities in the natures of these entities that make their comparison worthwhile.

First, let's consider whether the Holy See is locationless. It's true that its primary home is the Vatican City State, with its city-park–sized parcel of land in Rome, but it would make little practical difference if this location were merely the Holy See's embassy in Italy. The Holy See remains a formally distinct entity with an essential nature that is not tethered to a specific place. And although it possesses properties all over the world, it does not hold any significant habitable land.

Today, the Holy See has a global influence that has little to do with the existence of its real estate. Diplomatically, it is broadly recognized as being equivalent to a sovereign state and is represented by ambassadors in many countries. The Holy See is closely affiliated with large groups of people whose members are citizens of a wide variety of nations—the approximately one billion Catholics worldwide. This affiliation, of course, greatly influences the relationship between the nations in which these citizens reside and the Holy See. Moreover, the organization has a vibrant international economy related to its global mission and activities.

Returning to the humem-state, the Holy See example imparts a feel for what I mean by an independent, locationless humem-state. Like the Holy See, a humem-state will be closely affiliated with people in nation-states, yet it will remain mostly independent from these states in terms of its charter, administration, and economy. Likewise, alpha-people and others with relationships with the humems will have a strong interest in the humem-state's welfare. Thus, they will induce their governments to interact favorably with the humem-government. In this sense, the humem-state will gain political influence in the people-states. Persons acting as humem-government advocates, or diplomats, will be analogous to national bishops and other Holy See diplomats who function as the church's representatives within countries, and who are often also citizens of those countries.

In present times, since the Holy See is not another "regular" nation-state competing for people's citizenships, there are generally few practical conflicts of interest. Individuals can simultaneously be good Catholics and good state citizens. Of course, the members of the church can make direct or indirect monetary contributions to the Holy See that may result in the transfer of money out of their countries of citizenship. But, these amounts are typically modest compared to the citizens' total economic activity, and thus these donations are not a significant concern for the people-states. On the contrary, many states encourage contributions to religious organizations through tax benefits. Comparable dynamics will govern the interactions among people, their governments, and humem-states.

Perhaps the most striking lesson from this comparison relates to the benefits derived from the humem-states' eventual independence from nation-states. The core constructs and doctrine of the Holy See are relatively immune to the events in specific nations containing substantial Catholic populations. Nation-states may come and go, they may dissolve by merging with others or by breaking into multiples states, and borders may change. However, the Holy See prevails—its fundamental nature and basic purpose endure for centuries. Such is the vision for the humem-states.

Finally, I will reiterate that I am not proposing the Holy See's internal method of governance as a model for the humem-government. Rather, as noted previously, humem-states' governance should be similar to those of good people-governments. For our purposes, the primary characteristics worthy of evaluation, and perhaps emulation, are the Holy See's consistency of long-term purpose, its locationless nature, its interactions with other states and the populations of other states, and possibly also some of its economic underpinnings.

This cursory description of the ideal humem-state is just the beginning of a much more detailed study. My intention here is to impart a very general sense of how this state may appear and function. In the spirit of our gradual discovery, we will continue to build on this basic model as we study additional humem manifestations and behaviors, starting with humem economics in the next chapter.

But first, let's examine the interim solution: a pragmatic and immediately achievable system that will provide the urgently needed habitat for the rapidly emerging proto-humems, while also establishing the foundations for the gradual development of the more complete humem-state.

The Foundation—a Humem-Protectorate of the People-State

The first stage toward the ideal humem-state is a dedicated organization within a people-state—typically in the same nation where the majority of the alpha-people live. Most of the barriers and risks to the humem establishment can be removed by utilizing recognized legal and organizational frameworks, and by delegating standard functions to existing, trusted institutions.

To maintain a clear vision (of the ideal humem-state) and a consistent mission (to accomplish this humem-state), it is essential to properly name the constructs. We can call this new organization the *proto-humem-state,* and its inhabitants, *proto-humem-citizens.* We can regard it as a protectorate state, or *humem-protectorate,* under the auspices of the hosting people-state. In this context, the prefix "proto" signifies that, during the early stages, both the state and its citizens will not possess the full rights and abilities of their future manifestations. As we proceed, we will continually test the aptness of this nomenclature. I would argue that once we have studied the character of these bodies in more detail, and have understood the interactions between the humem-institutions and those of the people-state, we will find that this terminology is intuitive; in a sense, it would be unnatural to use any other.

In practical terms, the initial humem organization should be an allowable and recognized construct within the hosting state. In many jurisdictions, a noncommercial, or nonprofit, organization probably makes the most sense. There are numerous examples of purpose-driven political, religious, scientific, and cultural institutions, which are comparable to the humem institution in some of their attributes. In the past, some have even been dedicated to the establishment of foreign states or governments. There have been governments in exile, or bodies committed to the advancement of some external (to the hosting state) nationalistic interest. As long as such organizations are not perceived as conflicting with the hosting state's interests, they are usually tolerated. If they are congruent with the hosting state's policies, they may even enjoy active state support.

As argued earlier, for the long-term, a typical nonprofit organization is not the ideal structure for the humem-state; for the initial stages, however, it appears that it is the best existing option in modern nation-states. Let's examine a proto-humem-state organization from two different perspectives: the hosting government's viewpoint, and the people's—in particular the alpha-people's—viewpoint.

The Hosting People-Government's Perspective

In the early stages of humemity, people-governments will most likely be oblivious to the far-reaching significance of the entity that has started growing from within. Thus, initially, from the national government's standpoint, the humem organization should appear similar to other noncommercial, long-term–purpose organizations that exist within the state's jurisdiction. The organization's charter, which is required for its creation and recognition, should be transcribed into everyday terms that are familiar to the people-state officers who authorize the establishment of these kinds of institutions. For these official purposes, the charter should depict the more immediate and mundane humem applications, since anything else would most likely be incomprehensible to the state representatives. For example, one practical function of the humem system is the management and preservation of the whole-life data of many individuals. This has immense potential social value for historical, genealogical, memorial, and many other cultural reasons. Moreover, such a purpose is comparable to those of many existing noncommercial organizations.

Over time, the people-state will come to view the proto-humem-state as a desirable economic entity. The proto-humem-government employs people, both directly and also via the acquisition of services and equipment from commercial companies. As we'll see in the next chapter, the humems also hold assets—initially under the guardianship of the proto-humem-state organization. The people-state benefits from these resources remaining within its jurisdiction. Furthermore, if the proto-humem-state is more attractive than alternatives that are embedded in other people-states, it may facilitate an immigration of humems originating in other countries. Such humems may arrive with their assets, which can create an influx of economic activity without incurring any of the associated overheads, costs, and risks related to the physical immigration of people. [48]

By virtue of these and many other advantages, the people-state has little to lose and much to gain by hosting the proto-humem-government, or humem-protectorate, within its jurisdiction. Technically, the proto-humem-state could almost as easily be established in some foreign jurisdiction while still providing very similar services for humems

---

[48] In present economic terms this can be viewed as the export of a new product, humem-citizenship, which results in an inflow of foreign capital.

originating in the home state. Therefore, ultimately, people-states cannot maintain monopolies over humem-states. For now at least, governments do not commonly assign monetary value to personal data, and thus governments do not generally regulate it. It follows that the *emigration* of this humem body component (the data) is not presently prohibited. The national governments may, though, attempt to limit the emigration of humems' monetary assets to foreign jurisdictions. Due to monetary globalization and the typically modest sums involved for now, such constraints are unlikely to pose insurmountable obstacles to humem emigration over the long run.

From the people-state's perspective, besides the economic aspect, there is another fundamental advantage of having the early humem-state hosted locally, namely, the issue of control, or compliance. Numerous conventions and laws regulate behavior relating to the public expression of data. These include minors' privacy, incitement, libel, copyright laws, obscenity, state security, and many others. All these have counterparts in humem manifestations. If the proto-humem-state is hosted in a foreign jurisdiction, it is not necessarily obliged to abide by the laws of the alpha-people's state. This could be perceived as a threat to the nation-state's conventions and could result in damaging antagonisms causing the nation-state to attempt to impede the functioning of the humems in the external humem-state.

By contrast, if the proto-humem-state is hosted locally as a protectorate, it is innately obliged to comply with the laws of the nation-state. This results in the people-state's increased perception of control and provides it with recourse in case of humem infringements. Consequently, all parties can benefit from an initial "local" establishment of the proto-humem-state.

Again, the long-term vision is that of an independent humem-state in which the humem-government determines its own policies. However, I believe that the initial strategy must be one of maximum accord—within reason—with current states. Such an approach removes most of the obstacles to the humem establishment and thus is, by far, the shortest and safest path to the foundation of a competent and trustworthy humem-state.

In the humem-protectorate's subservience to the laws of the hosting state, constraints, where they may exist, are mostly ones of contemporary public humem expression, and less related to the constitution of the humems' core character. In other words, like people, humems can contain almost anything within their memories and minds, but they must still comply with certain cultural and legal norms in their outward, or public, appearance. However, such restrictions are often transitory and do not seriously reduce the potential for future manifestations in accordance with developing needs, relevancies, and conventions. As before, a complementary way of thinking about the trade-offs of this strategy is to consider that with the people-state's protection and support also come limitations—the temporary constraints of the humem childhood. Furthermore, due to humems' potential longevity, as long as they reside on solid foundations, they can be more patient with regard to their aspirations for freer future expression.

Presently, it is difficult to foresee the duration of the protectorate stage, the length of the childhood of humemity, or when a transition to an independent form of governance will result naturally. If the childhood persists long enough, it is conceivable that certain people-states may progress sufficiently toward humem concordance, thereby negating most of the objections stated earlier to the adaptation of people-state systems. In such an eventuality, the split into separate states may become redundant. However, in the light of most people-states' historical resistance to change, it appears that this is not very likely. Even if this turns out to be the case, in the interim, the initial solution of the humem-protectorate would still make sense for all the same reasons stated above. Insofar as such an adaption of the people-states may be possible, the proto-humem-states' early partial autonomies and their improving prospects for secession would provide the ideal impetuses for such people-state revisions.

Humems are the ultimate world citizens: they don't need roads, schools, hospitals, or land. They are not typically burdened by physical property. Like migratory birds, if the habitat is conducive to their needs, they'll come and stay; If not, they'll go elsewhere.

Thus, as humems mature and generate more value, the benefits of accommodating humems—i.e. being humem-friendly—will become increasingly obvious to people-states. This, in turn, will accelerate the development of the necessary humem infrastructures within people-states.

The Alpha-People's Perspective

As humems grow and develop, and we get to know them better, they will become precious to us—almost like children. We will only entrust their welfare to a system that can demonstrate a high level of integrity and competence. We have already seen that a commercial company is not qualified for such a role. Likewise, any untested noncommercial enterprises or foreign institutions will also be unacceptable. As alpha-people, we will look for the most solid and trustworthy administration available. Such criteria will strongly outweigh other considerations, like transitory limitations in the services that the administration provides. Even if, somehow, an independent humem-state emerges during the early stages of humemity, most of us will be justifiably apprehensive to entrust our alpha-humems with their associated assets to an innovative but unproven body. Unless an organization has the oversight and guarantees of more recognized and credible institutions, it will be difficult to convince us of its ability to achieve the long-term humem goals.

Embedding the proto-humem-state within a stable and trusted jurisdiction addresses most of these concerns. Thereby, the organization can acquire many of the assurances from the mature legal and cultural environment. A number of the crucial functions, such as the management of the humems' monetary assets, can be delegated to existing trusted institutions. Methods such as bidding processes for the selection of vendors performing the various technical services can emulate robust and transparent governmental procedures. Also, external, commercial auditing services can be employed very early on to emulate governmental checks and balances mechanisms. Devices such as these, together with the establishment of the humem organization within the same judicial environment as the alpha-people, can impart the trust and credibility that are so vital for the humem-state's preliminary stages.

Cultural, political, and economic familiarity will increase this early solution's appeal. A local proto-humem-state can most naturally clone and emulate the well-established behavior of the hosting state. Thereby, the alpha-people can more intuitively understand the processes and terminologies. They can recognize the way things are done because most practices have counterparts in their own people-state. The people, humems, and legal and financial environments are culturally harmonious. Everyone feels at home. The "proximity" of the embedded proto-humem-state also increases the people's perception of control. The familiar legal and economic environment facilitates easier redress in case of contingencies. In short, as the proto-humem-state takes shape, the nascent proto-humems—still in childhood—remain "close to home" and within easy reach of their alphas.

Granted, alternative organizations outside the people-state could provide similar services by creating customized interfaces for the alpha-people. The point is that an early humem-state, residing within a well-functioning and stable people-state containing a significant population of people who are related to its humem-citizens, will emanate far greater credibility than one hosted in a foreign jurisdiction.

This is not to say that humems from one jurisdiction will not prudently reside in foreign humem-states if they are superior to local options. Even early on, it is conceivable that certain nationals will not have confidence in the judicial or cultural environments of their own countries, and therefore may choose to install their alpha-humems in a proto-humem-state that is hosted within a more trusted jurisdiction. Under such circumstances, the alpha-person would still have more confidence in an established environment within a credible people-state, as opposed to a completely new and unregulated institution. Such has been the case for monetary assets over the centuries. Those who have been unsure of the stability or trustworthiness of their country or its financial institutions have often opted to entrust their assets to foreign institutions. Of course, similar actions are often motivated by other reasons, such as tax optimization. Comparable factors may also play a part in future humem-related considerations.

The Path—the Protectorate as a Locationless Humem-Colony

The proto-humem-state as a protectorate of a people-state is a conservative and solid strategy for the establishment of humemity.

An intriguing refinement to this concept of the humem-protectorate as an interim stage developing toward the independent humem-state is to think of it as a "virtuous" type of imperial colony. (It would probably be the only one of its kind.) The humem-colony can be considered virtuous in that it colonizes a truly virgin habitat without the displacement of indigenous peoples or contentions with other empires.

The colonies of the British Empire that resulted in independent states included the North American colonies, New Zealand, and Australia.[49] These colonies spent extended periods both enjoying the protections and bearing the restrictions of the mother country. In their beginnings, they did not technologically, culturally, ideologically, or legally need to "reinvent the wheel" for most purposes. All the fundamentals were cloned from the mother country. At different stages, and to various degrees, they gained their footing, developed their own character, and diverged from Britain. The process was gradual—sometimes with the encouragement of the mother state and other times through violent conflict. In the early stages, it was impossible to gauge how long independence would take. Even now, their roots and cultures are closely intertwined with those of their originator. In practical terms, it is hard to imagine how it could have been possible or advantageous for these colonies to have spawned from Britain and to have obtained autonomy in a much shorter period than they did—say,

---

[49] I selected these examples because they resulted in colonizations that (tragically for the indigenous people) were only marginally influenced by preexisting populations. That is, the resultant predominant cultures in these countries are much closer to British culture than to those of the indigenous populations. Thus, the actual outcomes in these instances approximate the colonization of virgin (unpopulated) lands and thus are analogous to the humem-state scenario. These examples stand in obvious contrast, for example, to the British Asian and African colonies where the indigenous populations and cultures eventually prevailed.

in five years, instead of, for example, the century and a half between the start of the first substantial North American English settlements and the independence of the USA.

I do not think that the historical periods between the conception of these colonies and their eventual independence are necessarily indicative of the time-to-independence of the humem-states. Due to the strong humem dependence on rapidly advancing technology, it is feasible that the humem-states may achieve autonomy much earlier. On the other hand, humem-individuals, unlike the first educated and able settlers of the English colonies, are still immature; thus, their preparation for independence may conceivably take even longer than we might anticipate.

As we proceed to examine the early humem environment in more detail, we will return to this conceptual framework of the locationless colony, and find that, in many instances, this analogy provides a consistent and intuitive model of the workings of the early humem-state and its interactions with the hosting people-state. I suggest that, in a sense, we may come to see this approach to the practicalities of the establishment of humemity as the obvious path forward.

Governments Regulating Humem Behavior

Previously, before examining governance in any depth, we discussed humem relationships. We saw that humems will emulate how people interact and they will form corresponding relationships. And, as is true for people, these connections will establish the framework for much of humems' behavior. This pertains both to interpersonal relationships and to interactions between individuals and institutions.

We also observed the alpha-pair bond in some detail as a predominantly new type of relationship—one that does not have a direct counterpart in traditional interpersonal associations. Now, having obtained a sense of the possible modes of humem governance, let's reexamine relationships in this broader context.

We have seen a number of reasons why, to achieve permanence, humems require eventual emancipation within dedicated humem-states. They need to be free from the constraints of people's location-centricity and mortality, and thus must be dissociated from their alpha-people's nation-state citizenships. If this is the case, an immediate question arises: If humems are not formally owned and controlled by their alphas, how can their conduct conform to their alphas' aspirations? In other words, how can we ensure that humems benefit people even though they are independent citizens of separate humem-states?

Alpha-humems are born of people, they are an extension of people, and the initial humem-state is a protectorate of the people-state. Therefore, humems individually and collectively are required to adhere to overriding principles that ensure the well-being of people, especially those who are their closest relatives. As I previously asserted in the discussion on "Humem Ethics and Obligations," humems must be *loyal*. This is largely in the humems' own interest; otherwise they will not be created or nurtured by people.[50] Moreover, if they are harmful, people and people-institutions will actively oppose them.

A devotion to these principles is a fundamental duty of the humem-state, which functions as the humems' guardian during their childhood and is therefore responsible for regulating their behavior and managing their "education." Humem behavior that conflicts with the wishes and norms of their human relatives is the single most dire risk to humem welfare, especially during the initial stages of humemity. Accordingly, the humem-government's obligation to its citizens, in many respects, entails an obligation to the alpha-people, since one cannot succeed without the other.

If a nation-state permits its citizens to aggressively confront others within the protections of its borders and does not enforce law and order to counter such actions, then typically the state is held accountable. Similarly, when humems develop into more autonomous citizens within an independent humem-state, the humem-state will continue to be

---

[50] In this limited context, humems are loosely analogous to domestic pets—nicer poodles result in more poodles.

responsible for enforcing the laws and guiding the ethics that regulate the mature humem-citizens' conduct, including internal and international interactions.

While the humem-state's oversight is vital for individual humems' success, it is no less crucial for the humem-state's survival. In the absence of such regulation, the people relatives would become dissatisfied. Consequently, new humems would not be established in the state, and existing humems would emigrate. In extreme cases, the humem-state would be sanctioned by other states and could become a failed state. In many ways, the dynamics in such circumstances would be similar to those of failed nation-states. Nevertheless, there is a least one key difference: Defunct people governments and states can sometimes persist for extended periods because they control the land on which their people live and thereby retain the resources necessary for the state's survival. Humem-states, however, will have few such natural resources; if a state deteriorates, its lifeblood, contained in its irrepressible humem-citizens, will be rapidly depleted as the humems easily relocate to more favorable jurisdictions.

CHAPTER 8

ECONOMY AND HUMEM AFFLUENCE

Providing sustenance for our survival is the fundamental motive for our economic systems. Likewise, the formation of a sustainable and dedicated economy is essential for humems' well-being.

Humems can be supported and subsidized in childhood by their alpha-people and other benefactors. However, in order to reach their full potential, they must eventually achieve economic independence. The sooner a humem becomes self-sufficient and legally insulated from its alpha, the less vulnerable it will be to the consequences of its alpha's eventual decline. Like a maturing human child, a self-reliant and capable alpha-humem has the strongest chance of achieving its highest aspirations. Moreover, independence makes a humem more valuable to others and more capable of assisting its alpha if and when the need arises.

As we study humemity in more depth, we will discover that humem economics shares many features with human economics, in both its diversity and its complexity. In this chapter, I will try to convey a basic sense of the core dynamics of humem economics and how it might unfold in the future. We will look at cases in which humem economic systems are similar to those of people, as well as cases that demonstrate key differences between the human and humem economic models. After that we will examine a number of contingency scenarios—in particular those relating to humems' prospects for survival in circumstances of economic downturn or catastrophic loss.

As we have until now, we will focus our discussion on the alpha-humem case. While other types of humems may exhibit different fiscal features, the alpha-pair exemplar is sufficient to demonstrate the fundamental concepts.

Initially, there are four primary stakeholders in the alpha-pair economic setting: the alpha-person, the alpha-humem, the humem-state, and the hosting people-state. Let's examine each in turn and consider the interactions and overlaps among these participants in the emerging humem enterprise.

Alpha-Person Economics

The establishment of humemity will be largely driven by alpha-people. Naturally, one of their foremost concerns will be their alpha-humems' long-term economic viability.

Especially in its initial stages, a robust and stable humem economic system will depend on how effectively the humem-state employs reliable financial custodians to manage the humems' assets and thereby ensure their fiscal health. During the humem-state's formative years, before its financial institutions have demonstrated their ability to achieve these goals, the economic risks and alpha-people's concerns will be especially acute. As I proposed earlier, the principal remedy for these inception challenges is to enlist the services and oversight of established institutions, such as reputable banks, insurance companies, and legal trustees. These can provide the necessary infrastructure and assurances until humem institutions can establish their credibility. This early-stage infrastructure is crucial for attaining humem economic resiliency as well as for alleviating the doubts and hesitancies of alpha-people.

Central to the humem system is a notion that may initially seem peculiar to many of us, namely the necessity to relinquish the formal ownership of "our" alpha-humems. In our discussion of humem rights, we saw some of the many disadvantages of treating humems like property. Such a status exposes them to numerous vulnerabilities, such as damage or seizure, both during the alpha-people's lives and after their deaths. Humems' release from such bondage is essential if they are to become everything we want and need them to be. In fact, this detachment of ownership must occur even in the interim proto-humem-state if early humems are to have good prospects for a long and prosperous future.

At first glance, this may seem counter-intuitive. One might ask, I create it, I care for it, and I pay for it...how can it not belong to me? Yet, all that is needed is a simple conceptual switch to resolve this quandary. If we think of our alpha-humem like a child, instead of like property, then the concept of ownership suddenly seems wrong and out of place. When speaking of a child, the "I create it, and care and pay for it, but it does not belong to me!" is entirely reasonable and familiar. People's motivations to invest in their alphas are similar to those of parents who provide for a child: they invest in perpetuity by first empowering and then releasing the carrier. As with one's child, if one clings too tightly to one's humem, it will fail to thrive and endure.

Frequently, ownership is a prerequisite for control. The alpha-person's waiver of possession, then, is not only a matter of finances but also one of authority. Here too, the child analogy can partially reconcile the conceptual problem of a humem that is not owned by its alpha, but which should, nevertheless, be congruous with the alpha's interests. Instead of a controlled humem, we can think of a devoted or loyal humem—one that is sensitive to the alpha-person's needs and desires.

While the child analogy may help us understand the need for humem independence and why humems cannot be owned by us, there is a notable gap in this parallel. That is, despite our good relationships, our children often have interests that are not absolutely aligned with ours. So instead of comparing the humem to a child, let's imagine the coveted alpha-humem as the most devoted friend conceivable—a soul mate whose basic aspirations are closely aligned with the person's. This humem friend, unlike the person's children, relatives, and human friends, exists in a domain that is extrinsic and complementary to that of the person, and thus does not have competing interests with those of the person. Such is the ideal nature of the alpha-humems that the humem-state must cultivate. In essence, the realization of these characteristics is the lifeblood of humems and the humem-state. Failing this, alpha-people will create substitute humems in other states, or they will not create them at all.

A valuable entity that is not owned by us but is nevertheless dedicated to us! Achieving such a construct may be challenging in our present jurisdictions due to regulations relating to asset ownership and taxation. Yet, if our current laws constrain how such entities may materially benefit us, then so be it—until humemity matures. Such temporary limitations do not fundamentally alter the long-term prospects for the continuance and autonomy of these entities. Over the long run, it is impossible to prevent humems from being established in other humem-states. Consequently, to discourage their relocation, existing jurisdictions will eventually need to adapt and become more conducive to humem needs.

Many people already allocate time, money, and other resources to objectives that are harmonious with humems' existence. For example, personal data retention and consolidation applications, and data security services are growing and increasingly being monetized. Humems will fulfill these functions and many more. Once it becomes evident that the economic efficiency of the humem system is superior to partial and localized related applications, then humemity's appeal will rapidly grow. That is to say, once people can obtain the humem solution with both its current usefulness and its alluring future potential for the same or lower cost than the temporary and disjointed related applications, then humemity will flourish.

Alpha-Humem Economics

Since the economic activity of mature humems will probably be as diverse as that of people, I will not attempt to describe it in any detail. Nonetheless, since the economic environment's early design should support aspirations for the future, we need to at least understand its basic tenets.

In this vision of the future humem-state, humems will possess economic characteristics that are similar to and will, to a degree, mirror those of people. Accordingly, their economic behavior will differ significantly between their childhood and their adulthood. As humemity matures, adult humems[51] will develop into autonomous economic entities—as adult people in people-states do.

Let's start by considering how affluence and cost of living can be understood from an alpha-humem perspective. We'll examine the first stages of humem economics, those of the childhood of humemity. Then we'll consider ways in which humems may produce income more dynamically. Finally, in light of the more complete humem vision, we'll look at how the alpha-pair's economics might be expressed as a form of partnership between the person and humem.

Affluence, Standard of Living, and Costs of Basic Sustenance

So much of humem economics is intuitive and comparable to traditional economics that the discussion may sometimes seem redundant. Yet, it is precisely this familiarity with our own system that can lead us to make erroneous assumptions in situations where humem economic behavior does actually diverge from that of people. In this section, I'll focus on some fundamental humem economic characteristics while emphasizing both the similarities and the disparities between humems and humans in this regard.

Economically, humems have varying levels of affluence, which allows them to achieve diverse *standards of living,* or if you prefer, *standards of existence.* With finite resources at their disposal, humems, like people, prioritize their expenditures. With low levels of affluence (in tough times), they contract and maintain a core. And in times of plenty (in good times), they develop, expand, and explore new possibilities.

However, there is a fundamental difference between humems and people in how affluence, or the possession of essential resources,

---

[51] In this context, adult humems are typically the alphas of adult people or they are ancestral humems.

relates to survival. Under harsh conditions there is a threshold beyond which people cannot survive; having crossed that boundary, there is no return. This is not so for humems. Something of a humem can survive under almost any circumstances and at almost every scale—there is no sharp delineation between life and death, or between existence and termination. In desperate conditions, humems have the capacity to "hibernate." A humem can contract to a small fraction of its full potential manifestation and maintain a minimal level of energy expenditure without terminating, or dying. Like seeds of a tree or shrimp eggs in the dry mud of a desert lake, humems are able to survive long periods of drought and then revive when the rains return.[52] Moreover, even if hibernating humems lose some of their core constituents, numerous avenues for revitalization remain. Since humems are physically dispersed and social entities, when the environment again becomes conducive to their sustenance, it is typically possible to regenerate parts of a humem from those of other humems, especially its relatives, as well as from fragments cloned from the global cloud of the collective EP.[53]

Most basic parameters relating to humans' physical existence have altered little over history. Our size and energy requirements (e.g., minimum daily caloric intake) remain fairly constant. It appears that our cognitive and physical abilities, on average, have also not changed very significantly. With humems, however, analogous measures vary considerably. For instance, in loose comparative terms, the energy, or financial resources, required to maintain an abiotic-EP of a given dimension, measured as information content and processing power, has been decreasing by orders of magnitude every few years.

This economic attribute has profound positive implications for the

---

[52] This analogy does, of course, break down at some point since seeds and eggs are ultimately discrete units that may entirely lose their viability under certain conditions.

[53] By global cloud of the collective EP, I allude to publicly accessible EP that is not necessarily clearly affiliated with a specific person or humem. An example is a written description or a photograph of a public event in which the alpha-pair participated but whose author is not necessarily known or particularly relevant.

long-term future of individual humems; it implies that the humem economics model is inherently robust. This essentially means that if a humem is economically viable today, then its prospects will almost certainly improve in the future! Put differently, we may say that, for a fixed cost, humems' standard of living is dramatically increasing over time. And for a constant standard of living, or even one that is gradually improving, the cost is markedly decreasing, i.e., the cost of basic humem sustenance is constantly declining.

In my experience, many people's doubts regarding the prospects of the humem-state focus on its economic feasibility. This is especially true for professionals pondering its practical implementation. To further address these legitimate concerns, later in this chapter, we will examine additional facets of the humem system's economic reliability and survivability.

Passive Humem Economics—Childhood and the Guardianship of Assets

Like that of a person, the humem's childhood is characterized by its inability to manage assets and the corresponding restrictions on its ownership of property. As we saw earlier, the alpha-person and the humem-state typically maintain joint accountability for the young humem's welfare. However, the humem-state bears the ultimate responsibility because the person may not always be available or allowed to fulfill this function. During the proto-humem-state's early phases, people may be legally restricted in their economic influence over their alphas, especially regarding their humems' monetary assets. (This is, nevertheless, consistent with the alpha-people's more general waiver of the ownership of their alphas.)[54]

So, at least initially, the proto-humem-state in its role as guardian must lead the management and safeguarding of the humems' assets and economic activities. Typically, the alpha-person will establish their

---

[54] Some existing laws relating to trusts, for example, dictate that people are not permitted to control funds that they do not own even if they are the original contributors to those funds.

alpha-humem under the formal control and ownership of the proto-humem-state, while the humem-government will be responsible for inducing the humem to adhere to the person's best interests wherever possible. As appropriate and necessary, during the person's income-earning lifetime, the person will provide the humem with ongoing sustenance and gradually establish an economic foundation for the humem's long-term needs. Beyond the person's life, the humem-state will continue to maintain and grow the humem's assets and economic integrity until such a time as the humem is able to perform some of these functions more autonomously.

In the simplest and most conservative scenario, one that is probably typical of the early stages of humemity, the alpha-humems' assets will consist of various forms of perpetual funds, annuities, pensions, and so on. These monetary instruments will be developed by the humem-states and will be adapted to the humem population's needs within the frameworks of the hosting nation-state jurisdictions. We can refer to these types of economies, in which the people provide for the humems, and the humems subsist from the proceeds of investments, as *dependent* or *passive* humem economies.

Dynamic Humem Economics—Humem Livelihoods

Following the initial establishment of humemity, we can expect a multitude of additional humem economic manifestations. Since the humems are an extension and a mirror of their alpha-people, humems' economic activities will increasingly converge with those of their human originators. Thus, to advance humemity, it will be imperative to allow humems more economic expression as soon as it is feasible. Again, if such opportunities are denied by a particular humem-state or obstructed by a hosting people-state jurisdiction, they will eventually emerge somewhere else and result in the migration of the affected humems to environments more amicable to their needs.

Economic freedom does not necessarily imply economic expertise; humems, like people, can benefit from the skills of capable asset managers and other related service providers. In the beginning, these functions will most likely be performed by the humem-state's financial management services, but later the humems should have a broader range of options for their financial administration needs.

In contrast to the passive kinds of economic means described in the previous section, a *dynamic* humem economy is one in which the humems take a more proactive role in generating income. In our growing information economies, various manifestations of our EPs already play a big part in our professional interactions and have far-reaching implications for our productivity. We are also seeing the beginnings of direct revenue generation by proto-humems and other concentrations of our EPs.

For example, there are many personal EP expressions, or appearances, for which people pay to gain access. At the present time, these most commonly take the form of books, music, movies, and educational or entertainment websites. Often, they are aggregations of people's EPs and reflections of at least part of their characters; thus, in accordance with our new terminology, it is consistent to call them proto-humems. Such an entity's income is generated in various ways: it may come from direct customer payment (for interaction with the proto-humem), or it may be derived from advertising revenue (from ads placed by third parties in conjunction with the proto-humem). In both these examples the proto-humem's intrinsic value is the source of the revenue.

An early illustration of an income-generating proto-humem is an instructional Web site commonly depicting an instructor—or more precisely, their EP as an emulation of their personality and character—communicating by means of text, images, video, and interactive chat. Often, the learners, or subscribers, can request clarifications and additional material, undergo progress assessments, and interact with each other. This situation exemplifies a direct and multifaceted interaction between people and a proto-humem. In present settings, most of the humem's more complex responses to the learners' input, such as answers to questions, are performed by the alpha-person. However, some basic answers may already be supplied by the humem; when the learner submits a query, the proto-humem may reply with something as simple as: "Thanks for posting your question—I'll reply within two days."[55] This is akin to a human administrative assistant providing a standard response before referring an issue to their manager. Like the administrative assistant, as the humem gains experience and ability, it will perform increasingly complex tasks more independently. In current nascent implementations, automated online assistants employed by some major corporations, such as utility companies, are addressing a broad variety of customer inquiries. As the related technologies advance, they will inevitably become more widely available to and utilized by the proto-alpha-humems.

Another type of existing, revenue-producing proto-humem application, typically for adult entertainment purposes, is the proto-alpha-humem of a charming or physically alluring person. This humem is typically portrayed via a Web site expressing various facets of their intimate EP using similar media to those described in the previous examples. The visitors interact with an entity—the proto-humem—that emulates the person's appearance, character, emotions, and intimate behavior. If the proto-humem is particularly engaging, visitors may pay directly for the interaction, or third parties may pay to have their less appealing content—such as advertisements—adjacent to the more appealing humem.

Notably, the visitors, or clients, can often affect these proto-humems' characters. For example, the visitors may leave comments or upload additional content, thereby influencing or augmenting the proto-humems. Such humem enrichment, which may in turn be experienced by future visitors, often occurs without the alpha-people's direct intervention. In a currently rudimentary manner, these dynamics mimic traditional interpersonal exchanges, in which people's minds are influenced and, to a degree, changed by their dealings with others.

Another simple example of a proto-humem's income-generating capacity is the circumstance in which a humem is assigned the ownership of an electronic book written by its alpha-person. By selling the book via an appropriate channel, the proto-humem may gain the means

---

[55] We may pause and ponder who exactly is the "I" in this response?

of income indefinitely (or for as long as the book remains relevant and sellable).

Perhaps most significantly, numerous commercial companies are already demonstrating the intrinsic dynamic economic value of the common humem. At the time of this writing, some of the world's most valuable online companies are deriving immense value from their collections of subscribers' data and associated applications—the subscribers' proto-humems. The fact that these proto-humems mostly exist without direct monetary support from their alpha-people suggests that, on average, they are able to generate enough income to be self-sustaining. Moreover, as evidenced by the healthy profits of many of these companies, it appears that the proto-humems are able to generate far more revenue than what is required for their current level of existence.

Now some people may still object, saying that a machine-based humem cannot be considered a self-sufficient income-generating entity. However useful it may be, it is just a device owned by a person or an organization. Yet, the same could have been said about slaves. For example, in places where slavery was institutionalized, renting out a slave was equivalent to renting out a horse or plow in commercial and legal terms. As with slaves, once humems achieve emancipation and become recognized, autonomous entities with the rights to own property and sell their services, then everything changes. Humems will have the ability to earn their keep and perhaps a lot more. Like slaves, the proto-humems' current economic situation comes at a grave cost to their dignity and self-determination, and to their ability to fully enjoy the fruits of the value they produce. A dedicated humem administration within a humem-state can direct this value to benefit humems and to advance the humem ecosystem in which they exist.

Currently, in most jurisdictions, the formal allocation of autonomous ownership to proto-humems may present a practical challenge. Nonetheless, as with many future humem rights and capabilities, these limitations will eventually prove to be transient obstacles that will be overcome as soon as there are enough incentives to do so. In the interim—during humemity's childhood—the proto-humem-states can bridge many of these gaps. These administrators, as guardians of the humems' assets, can maintain their formal ownership and control until the humems' legal and financial environment matures sufficiently to facilitate the transfer of possession and responsibility to the rightful owners. Continuing our metaphor of the humem childhood: at that point in time, the humems would have reached the age of majority, or "come of age," in humemity.

The Alpha-Pair as Economic Partners

In our description of a humem's economic development, we first envisioned the alpha-person, or a representative such as the alpha-person's parent, establishing the humem's financial foundations. Thereafter, the person would continue to support the humem until enough resources, such as investments, were amassed to sustain the humem indefinitely. These scenarios portray a relationship of dependence between the person and humem. We then considered ways in which humems, as extensions and emulations of people, might participate in a broader variety of income-generating endeavors while gradually achieving greater financial independence.[56]

If and when these more ambitious expectations materialize, and a dynamic humem economic system in which humems generate revenue becomes prevalent, there is a more interesting and satisfying way of viewing the alpha-pair as a joint economic entity. In such circumstances,

---

[56] The correlation between a humem's and its alpha-person's affluence is comparable to that between people and their parents. A child's affluence is strongly derived from that of their parents. In historical times, it also mostly stayed constant into adulthood: peasants' offspring remained peasants and the descendants of nobility retained their parents' privileges. This continuity of socioeconomic status over generations is still the norm in undeveloped countries today. Even in modern developed societies there is still a strong statistical correlation between a person's affluence and their parents'. However, especially in modern democracies, dramatic divergences are much more common—the offspring of the poor can sometimes achieve great financial success. Similarly, during a humem's childhood and during the early stages of humemity, it is likely that a humem's affluence will be strongly linked to that of its alpha-person. Later, independent humems in more advanced humem environments may be able to achieve far greater prosperity than in their youth.

we can consider the alpha-pair as a financial unity or partnership in which both members have compatible attributes and aspirations. As children, the alpha-person and alpha-humem are both sustained by parents; as they come of age, they become an economically autonomous pair; thereafter, they generate income together and share expenses while being motivated by common goals.

Early in these types of partnerships, there may be concerns relating to the division of resources between the person and humem. The person will still utilize the bulk of the proceeds for their traditional expenditures, and the alpha-humem will be assigned the amounts needed for its sustenance. Still, apparent conflicts of interest may arise, mainly due to the lack of knowledge about whether current resource allotments adequately address humems' long-term needs. In the beginning, this process may resemble the allocation of budgets between the various departments of an organization. While the overall goals of the organization should rationally guide such decisions, in practice there can be considerable discord that requires ongoing management. As the systems mature, these allocations should become more like the sharing within a harmonious family. Ultimately, when the alpha-pair becomes even more closely integrated, the distribution of resources within the alpha-pair should resemble the distribution of energy within a single living organism. It should become an automated and streamlined process, which adapts to fluctuating internal needs and a variable environment.

When viewing today's dependent "baby" proto-humems, many people may view this vision with some skepticism. I would argue, however, that we require a fundamental change in mind-set to properly envision humems' future economic nature. Already, in modern societies, many of our interactions with the world occur via our abiotic-EP. With the consolidation of the personal EP into humems, much closer emulations of our abilities will become possible. Once humems are able to take on some of these roles, and if these are income-generating functions, then they will create revenue. Moreover, humems will be able to surpass our capacities in many ways: they do not tire, they can exist in multiple concurrent instances, and they can manifest anywhere at any time.

Long before humemity achieves such advancements, however, it will become clear that the establishment of robust and capable humems will, above all, be a prudent investment for the alpha-people. Investing in the past, the present, and the future will all be transformed by the emergence of economically viable humems.

Humem-State Fiscal Fundaments

Humems and people's economic similarities are reflected in the economics of the humem-state. The concepts and structures are mostly familiar. Humems will pay taxes—probably as a function of their wealth and perhaps other factors such as their seniority. The state will subsist on this income and on the proceeds of other investments and assets under its control. Like nation-states, the humem-state will have ample opportunity to leverage its exclusive position and to monetize the services it provides. The resulting revenue will help finance its operations and ideally will also be reinvested in further humem development.

But, during the formation of the state, some things will be markedly different. For one, the guardianship of early humems will require special mechanisms. The state tax may initially take the form of humem asset-management fees, paid out of the humems' funds, or alternatively, guardianship service subscriptions, paid by the humems' creators. There are, of course, numerous other modes in which this could play out. With enough motivation, fiscal solutions can be adapted to the situations at hand.

In the beginning, some humem-governmental behaviors may be reminiscent of certain socialist states in which the government assumes close responsibilities for the welfare of individuals. For example, the guardianship could include components that are analogous to a state pension or national healthcare plan.

Yet, in other ways, the relationship between humem-citizens and the humem-state will be very businesslike and competitive. As we have seen, the equivalent of emigration, or relocation, to more advantageous humem-states will be much simpler for humems than for people. The state's economic appeal will be a key factor in the alpha-pair's choice of humem jurisdiction. Furthermore, the state's financial policy with regard to emigration will, in itself, be a crucial consideration. If a humem-state has policies that hinder emigration with assets, we will refrain from establishing humems there in the first place.

Consequently, humem-states will need to be economically efficient, innovative, and—above all—trustworthy in order to prevail against the competition of other states and to gain and retain satisfied and confident humem-citizens.

Independent nation-states typically establish and regulate their own monetary systems. As a protectorate, the proto-humem-state will probably adopt the hosting people-state's currency. As it matures, however, it may require its own currency. Intriguingly, we have recently seen the emergence of digital, or virtual, currency systems.[57] These decentralized systems are essentially locationless, like humemity, and the value they carry is linked to computer processing power, which produces a natural affinity between them and the humem system. While it is difficult to know how the humem monetary system will eventually develop, these similarities may provide some insight to the possibilities.

People-State Fiscal Benefits

In the previous chapter, we discussed some of the numerous ways in which people-states may benefit by hosting proto-humem-states. In particular, they can profit by facilitating the humem-states' economies from within their own systems. Humemity creates huge economic opportunities for people-states other than those gained from directly hosting the proto-humem-states. Humem-states will be affluent and insatiable consumers of technology and related services, many of which will be supplied by corporations within people-states including states other than the hosting people-state. Humem-states will employ people both directly and indirectly, irrespective of the people's locations or

---

[57] Also called cryptocurrencies.

nationalities. Humem systems will also require energy and other commodities, which will be procured from providers within people-states. Because of this economic interdependence, and in order to advance their economic stability, affluent humem-states will also probably invest heavily in people-state monetary instruments and other assets.

Historically, population growth has been an important component of economic development and has been an axiomatic factor in many financial models, such as those of pension funds. However, the recent stabilization of populations in several developed countries is requiring fundamental revisions to such approaches. In contrast to societies of people, the population of humemity does not have any upper bounds. Therefore, humemity opens the possibility of a new pattern of population growth: stable people populations coupled with constantly increasing humem populations. This implies that once humem-citizens obtain comprehensive economic rights and abilities, their presence may produce a continuously growing and potentially vast economic wellspring. Furthermore, unlike the growth of human populations, this expansion can be achieved without burdening the environment or depleting natural resources.

Economic Robustness and Contingencies

We have seen that economic viability is an essential element of the humem vision. The humem system, therefore, requires an economic system that can achieve stability over long periods with mechanisms to account for downturns and misfortune.

A well-structured, purpose-driven humem-state provides the foundation for realizing these aspirations. As in well-functioning nation-states, the underlying premise is that the vast majority of humems should be economically self-sufficient in the humem-state. Nevertheless, in real systems, especially over long periods, mistakes are made and calamities happen. So, let's now assess some safeguards and remedies against the more common kinds of economic contingencies.

Humem-State Social Security

Modern and affluent societies commonly have provisions for the care of destitute citizens. This is true for most advanced states irrespective of their forms of government or underlying ideologies. Notwithstanding the cultural reasons for this kind of benevolence, we can simply observe that such behavior exists and postulate that since humem-states will emulate nation-states, it is reasonable to expect similar conduct in humemity.

This prediction is further bolstered by the fact that the cost of humems' basic sustenance is much lower than people's. Humem jurisdictions that demonstrate such benevolence and include social-security provisions in their constitutions will be considerably more appealing to alpha-pairs deliberating their choice of humem abode.

There are various ways to implement humem social security, most of which have counterparts in traditional states. For instance, fiscally able humems may pay a form of social insurance, the rate of which may depend on their total worth. Graded insurance costs may result in different levels of coverage. Alternatively, the premium may be part of a basic humem-state tax with standard coverage for all.

Also, depending on the humem-government's policy, destitute humems may be required to participate in communal undertakings, such as historical or medical research, in exchange for the basic state support. This, of course, may raise ethical and privacy issues, as is sometimes the case for comparable situations in people-states.[58]

In presenting these examples, I do not intend to stipulate the appropriate behavior for humem-states in these regards. Rather, I wish to impart an awareness of the possibilities and the factors that will probably come into play in these future systems.

---

[58] In some countries, for example, patients receiving subsidized health care in state hospitals are afforded less privacy than paying patients in private hospitals.

Bailouts and Bolstering by Kin

Like people, humems have mutual affinities with family, friends, and others of the same tribe, ethnicity, spiritual belief, political affiliation, and so on. For our present purposes, let's use the term "relatives" in a broader sense to include all those individuals forming such relationships.

With humems the scope of kinship may be somewhat different from what we discern in ourselves and other species. While humems' kinships include members of their own kind, their kinships also extend to their alphas and other human relatives.

Generally speaking, we can say that individual humems and people are valued or esteemed by their relatives. In other words, relatives have an interest in the welfare of those to whom they are related. These empathies can act on many levels, and they often extend far beyond traditional familial or tribal relationships.

For instance, an ailing person most often receives assistance from close family or friends or clan, but in modern societies many other types of affiliations—such as religious associations, city residency, citizenship, and so on—come into play.

There are many explanations for this altruistic conduct. Some rationalizations are derived from cultural, religious, and ethical norms, while others stem from genetic and evolutionary considerations. But again, we need not go into any such analysis here; instead, we can merely observe this behavior and infer that whatever holds for us in these respects will most probably also be valid for humems. Moreover, there are reasons to speculate that, in certain ways, kinship may even be more valued in humemity. Due to the inevitability of people's decline, relatives may sometimes be hesitant to invest resources in trying to rehabilitate a seriously ill or elderly person. In humemity, however, the benefits of restoring a needy humem relative may often be much more tangible and enduring.

The prospect of a bailout, or bolstering, of an impoverished humem by its kin is not a distant prediction. People are already investing significant resources in a fledgling form of ancestral humem rescue.

Currently, the EPs of most people's ancestors are very partial, fragmented, and in constant decline. The remnants of these individuals' EPs are often held in dispersed shoeboxes or files as old documents or photographs, or as scant records in institutional archives. Even the EPs of more recent ancestors still exist mostly in the minds of the surviving people who knew them. Also, the social interconnections among the individuals' EPs have often disintegrated. In recent years, facilitated by technology, the price-performance ratio of genealogy services has improved sufficiently, resulting in millions of people investing time and money in a re-creation of their ancestry—the restoration and consolidation of their ancestors' EPs and interconnections. These developments are accelerating. Remarkably, we are already observing interplay between the biological relatedness of people and their abiotic-EP interconnectivity. For example, where documentation is lacking, genetic analysis is increasingly being used to discover and reestablish lost relationships.

Thus, the large-scale rescue of ancestral proto-alpha-humems by human descendants has already begun. In the future, in addition to this trend, capable and affluent humems will be able to assist needy humem relatives—ancestors, peers, and descendants. While the support may often consist of financial or equivalent resources, related humems also contain parts of the essence—the humem body—of their needy kin. They will be able to share memories, knowledge, and culture with their relatives.[59] They may thereby contribute in multiple ways to restoring their ailing relatives' good health.

Other Humem Benefactors

So far, we have focused on two types of safety mechanisms for humems: one provided by the humem-state, and the other provided by people and humem relatives. Extrapolating from what we see in the

---

[59] In a sense, we may call the reestablishment or reinvigoration of previously lost or degenerated humem information through transfer from another humem, *humem reminiscence.*

people-world, we can expect a number of additional factors to play a part in improving the long-term prospects of individual humems.

Various forms of commercial and mutual insurance instruments can be used for added protection above and beyond those provided by the humem-state. As we have seen, humems are less prone to absolute termination, or death. But they can be subject to reduction or degradation. Accordingly, they require insurance that is more analogous to health coverage than life coverage. With respect to humems' core abiotic constituents, it appears that, in principle, there are no fundamentally unavoidable or incurable maladies. Given enough motivation and resources, by using a variety of technical means such as data replication and encryption, the humems' health and longevity can be virtually guaranteed.

By properly structuring costs and coverage, wherein the expense of safeguarding a humem's well-being is significantly less than effecting an insurance payout, insurance providers can be strongly motivated to institute the procedures necessary to prevent such humem losses from ever occurring. Put in another way, the insurance could be calibrated so that the costs of maintaining humem health would be substantially lower than the costs of humem sickness. As a conceptual illustration, with permission and the appropriate provisions for privacy, etc., the insurance provider (or an authorized third party) could replicate the humem in its entirety, thereby ensuring its full reinstatement in case of any calamity.

Philanthropic, religious, and other non-governmental organizations often play significant roles in improving the situation of disadvantaged people. Similar organizations may provide comparable services for humems. Here too, such activities are already emerging. For instance, there are nonprofit organizations dedicated to archiving the content of the World Wide Web, which includes extensive EP and proto-humem material. The Internet Archive organization is well known in this capacity. Another salient example of an organization that perpetuates individuals' EPs is the Mormon Church, which has invested substantially in genealogical information, the retention and development of which is closely related to some of the church's core tenets.

Additionally, organizations are making various other efforts to retain life stories, family histories, and other cultural components for a variety of purposes. Since humems' essential makeup is rich in such elements, the motivations driving these undertakings often have a high level of congruence with those of humemity. This portends an optimistic future of cooperation and mutual support.

As we will see later when we discuss various other humem applications, humemity provides the ideal and unified platform for many of these currently disjoint efforts. Notably, since the humem system is an intrinsic carrier of family history and relationships, it is also an excellent genealogy device. (I would argue that there is no better method for maintaining extensive genealogy.) However, this same infrastructure is extremely versatile and supports many personal data applications besides familial interconnections and history.

Genealogy is but one exemplar of the mutual dependence and value that exist between and among humems and people. These reciprocities result in a form of mutual protection where the power of the collective promotes the welfare of its individuals (both human and humem). When independent individuals and organizations—with diverse underlying incentives—find a common goal in fostering humems, the prospects for humemity become much more secure.

Part IV

OUTCOMES AND OPPORTUNITIES

CHAPTER 9

FACETS OF HUMEM ANTHROPOLOGY

Now that we have established a basic understanding of the cultural, legal, political, and economic foundations of humemity, we can resume our examination of the varieties of humem expression.

First, we will extend our analysis of the implications of a lifelong-and-beyond alpha-pair continuity, and what this can bring even after the alpha-person is long gone. Then, we will examine a few of the many modes of humem creation and reproduction, and some of the tantalizing and imminently achievable related outcomes. Finally, we'll consider some variants of humems—ones that are very different in composition and behavior from the alpha-humems that we have studied until now.

In this book, and in this chapter in particular, we observe both general humem behaviors and more specific humem exemplars. This approach of interweaving generalizations with case studies, including descriptions of the behavior of individuals and groups and their modes of interaction with the outside world, is reminiscent of certain anthropological accounts of newly discovered societies. Thus, I find it natural to classify this inspection of humem origins and characteristics as *humem anthropology*.

To some, the aptness of this terminology may seem questionable: should it be just an unqualified "anthropology" as an expansion of the familiar term, or another term altogether? Among other considerations, this depends on how we will view humemity—as an assemblage of a new kind of being or simply as an extension of humanity.

Alpha-Alikeness and People's Age-Variance

Over the course of our lives, our personalities, physical attributes, knowledge, and behaviors change in many ways. For our present use, let's call these variations in our character over time our *age-variance*. In general, when comparing our natures at different stages of our lives, the age-variance increases as a function of the time elapsed between the stages.

As a humem's ability to emulate its alpha improves, in many kinds of expressions, the alpha-pair will become more alike. We can call this resemblance between the person and humem the *alpha-alikeness*.

Due to the person's age-variance, as the alpha-alikeness increases, the members of the alpha-pair will become more similar to each other, in many respects, at any given point in time than the person is to himself or herself at very different points in time.

For example, most adults cannot emulate themselves at age three with any accuracy. A capable future humem should be able to do this quite successfully. Another stark demonstration of this idea is the comparison between a lucid and healthy fifty-year-old, and the "same" person with dementia at the age of ninety. Here too, the humem at age fifty will be a much closer emulation of the person's preferred, or representative character, than the ninety-year-old human.[60]

---

[60] As we have seen, a humem will be able to "act" any age. So when we say "the humem at age fifty," we mean the humem as it appears when emulating, or representing, the fifty-year-old person.

Figure 5 schematically depicts this concept of alpha-alikeness and people's age-variance. With respect to certain characteristics, the commonality, or overlap, within the alpha-pair is greater at any given age than the overlap between the person's natures at significantly different ages during their life.

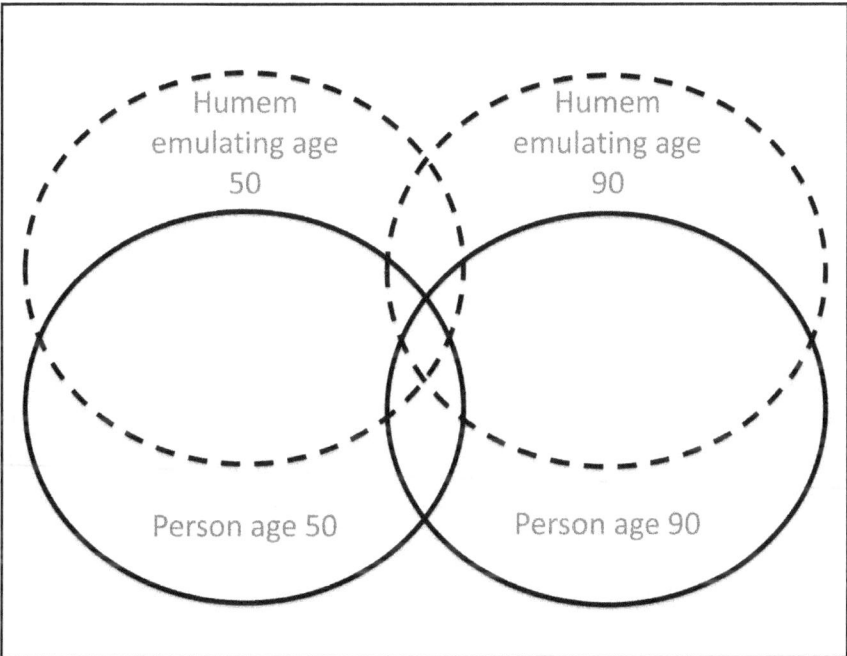

Figure 5: Alpha-alikeness and people's age-variance. With respect to certain characteristics, the commonality within the alpha-pair is greater at any given age than the overlap between the person's natures at significantly different ages.

The Alpha-Pair's Continuity over the Person's Decline

As alpha-humems develop and become more capable, they will become an inseparable part of our lives and how we communicate with the world. Furthermore, if they are designed and created to be mostly impervious to people's eventual demise, they will profoundly change both the implications and our perceptions of death.

It is useful to view the alpha-pair as a union of the bodily and extended presences—a kind of super organism—and to consider the nature of the whole over time.[61] Figure 6 shows the contributions of both the bodily presence (physical person) and the extended presence (humem) to the alpha-pair. The bodily presence grows from infancy, typically reaching a peak in capability and significance in mid-life. In the case of sudden death, as depicted in the lower diagram, the bodily presence can cease almost instantly. Nevertheless, in the modern world, a gradual physical and cognitive decline, as depicted in the upper diagram, is more common. A sadly widespread current example is the gradual departure of one's character as a result of Alzheimer's disease or other forms of progressive dementia.

When we consider the circumstance of a person's gradual mental deterioration and view the alpha-pair as a unit, nothing very remarkable changes in its interactions with the world at the moment of bodily death. Long before that time, the humem should be conducting most of the alpha-pair's communications and dealings with the world. In the diagram, the solid upper-curve, which represents the alpha-pair's total presence and influence, exhibits a fairly smooth continuum during this transition.

In existing systems, death is often accompanied by substantial legal and financial disturbances, especially for the deceased's dependents. However, a capable and independent humem should be mostly immune to this event. Legally and financially, it should be completely separate from the person. Accordingly, in most respects, the alpha-pair's character—outwardly at least—need not change much at this particular time. From an alpha-pair's perspective, the body's death should be no more than another milestone on its timeline into the future. While the sudden-death scenario somewhat reduces the seamless transition, in

---

[61] Later, I will introduce the notion of the alpha-pair as a unity called the *alpha-individual*. Initially, this can be thought of as an abstraction composed of two concrete entities, the person and humem. However, over time, the division may disappear with the alpha-individual emerging as the single definitive entity.

this case too, a well-established and independent humem should ensure continuance and carry on with its long-term prospects mostly intact.

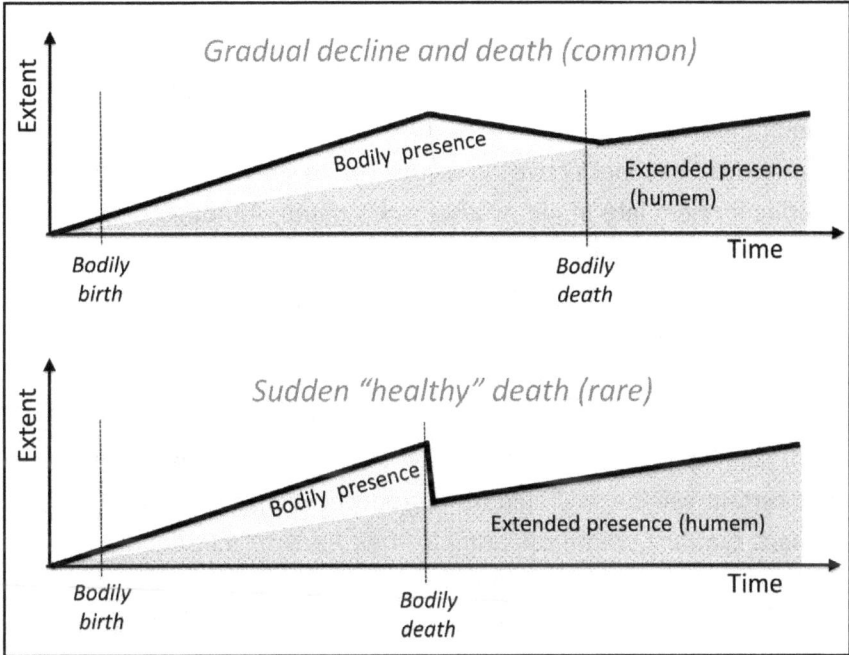

Figure 6: The alpha pair's continuity over the person's decline.

This depiction of the alpha-pair's continuum stands in stark contrast to the current reality, in which a person's EP in life is often detached from the EP of their afterlife. Such separations result in the decline and eventual disappearance of much of the individual's EP. Beyond their physical body, so much else is needlessly lost.

Many people expend considerable effort creating an afterlife EP for a loved one. Sometimes this kind of EP is expressed as a physical object, like a monument or plaque commemorating the departed person's life. These customary forms of personal heritage usually depict only a meager part of the person's richness of character, and often they do not endure for very long.

More contemporary attempts at creating legacy are also frequently inadequate. In recent times, for example, we see the emergence of online memorial, or legacy, applications wherein surviving family and friends create an afterlife EP by remixing some of the remaining fragments of the deceased's life. However, the outcome of these efforts is a representation of not the *deceased's memory* but rather other people's patchy *memories of the deceased.* On average, they do not endure for more than a few years—much less even than the traditional types of monuments.

Likewise, it is ineffective or impossible to create a comprehensive afterlife humem late in life or after one's death. Although one's alpha-humem is substantially different from one's child, we have seen parallels in the process of their creation and growth. This partial analogy can further elucidate the flaws inherent in delaying the creation of an afterlife humem.

Without the charity of others, for example, children could not develop or survive if we produced them at the end of our lives—in the way that certain salmon and spiders reproduce. Higher social animals and humans typically create offspring in their early to mid adulthoods—at the prime of the parents' ability and knowledge.

For creatures that mature primarily through social interaction, an extended coexistence with their parents is essential for their development and education. For similar reasons, humems require a lifelong coexistence with their alphas to achieve their full potential as the perpetuators of the alpha-pairs.

Now, departing from the imperfect analogy of a child, alpha-humems, freed from the constraints of biological procreation, can thrive and coexist with their alphas long before, and even longer after, the people's reproductive age. (We will consider some processes of alpha-pair formation in more detail in the next section.)

Once the alpha-pairs' continuity is secured by the humems' attainment of recognition, rights, and autonomy, the extent of individuals' influence will be fundamentally transformed. The fresh worldviews shaped by these new realities will have far-reaching implications for our perceptions of life and death.

Reproduction, Sex, and Instinct

Until now, we have observed alpha-humems as emulations, expansions, and in a sense, continuations of individual persons. Consequently, it is reasonable to ask whether humems will be able to emulate the most ancient form of human continuity, namely, sexual reproduction—a merging and partial continuation of two parents.

When contemplating how some counterpart of sexual procreation could apply to humems, it is immediately clear that many variations and processes are possible. We will touch on a few examples in this section—which, again, is intended to convey a broad appreciation of the possibilities in this area while neither prescribing nor predicting the dominance of any specific approach.

We have seen that alpha-humems can mirror many of our behavioral and cultural qualities. Likewise, they can gain a detailed knowledge of our bodies via wearable sensors or other methods. For example, one's humem can contain the results of one's genome sequencing—providing it with an increasingly meaningful insight into one's bodily makeup. Consequently, humems can emulate multiple facets of human beings—body, mind, and social behavior.

In traditional reproduction, over time, the offspring typically become the product of both the biological and cultural contributions of their parents. The question is, therefore, whether a merging of humems can emulate human procreation. In particular, can a coupling of parents' alpha-humems form the core of an appropriate alpha-humem for a human child?

What does "appropriate" mean in this context? Who decides? Well, in principle, these questions are not very different from those we might ask with regard to a child's physical makeup, personality, beliefs, and aspirations. Children do not get to pick their parents or their genetic makeup. Nor do they choose their native language, their teachers, their religion, and so on. In childhood, at least, and frequently thereafter, the majority of people adhere to the customs that the circumstances of their birth have thrust upon them. Thus, in practice, the nature of their parents and that of the surrounding culture largely determine what is "appropriate" for them. For example, if both parents have similar behaviors, beliefs, accents, or common physical attributes, then—once the surrounding culture is factored in—a lot can be inferred about their child's physical and behavioral characteristics. We can expect similar dynamics to direct the formation of children's alpha-humems.

Notably, the outcome of the humem coupling can be a lot more sophisticated than a simple average or random mix of the contributions of the parents' humems. The child's humem can be knowledgeable about its grandparents and other relatives, and possess an awareness of its surroundings and the progression of cultural change.[62] For instance, it can gain an understanding of the tendencies of cultural and physical deviations of other progeny from their parents in environments similar to its own. By superimposing its parents' contributions over the predictive models created from these trends of the younger generation, more relevant and appropriate "baseline" alpha-humems can be created.

Until now, we have mainly viewed the alpha-humem as *reflecting* the person's nature. For a child, especially, we can also conceive the humem as *directing* parts of the person's development. Earlier, we saw how humems could function as personalized mentors and guardians for children while augmenting the efforts of parents and other educators. The point here is that part of the makeup and knowledge of such a humem mentor can be inherited from the parents directly via their alpha-humems. A "newborn" alpha-humem can contain *inborn* capabilities, such as the ability to guide the person during infancy and in later life stages. Just as babies contain dormant aptitudes, such as parental instincts, which are only manifested much later in life, the baby humem can also contain abilities that may be usefully expressed or utilized in the future. An infant's alpha-humem could start developing with the birth of the child, or, as a better emulation of personhood, it could be

---

[62] To gain a more solid feel for this notion, a simple illustration may help: My children's computers, for example, already contain a sizable collection of digitized photographs from my spouse's and my childhoods. They also contain similar material depicting their grandparents and great-grandparents. That is, my children's nascent proto-humems already possess the rudiments of the merged memories of their paternal and maternal lineages.

established adjacent to the child's conception. By the time of birth, the humem could already possess a vast store of inherent knowledge and faculties, or *humem instinct*.

Publicly available information, such as that which is widely accessible from the global information base (say the Web), is less valuable as part of the new humem because it is typically not a distinctive part of the humem's character, and also because it will be readily available in the future anyway. However, information that is exclusive to the individual or to their kin is much more relevant. This may include, for instance, family history, the guidance of deceased ancestors, their genetic information and medical records, and even their culinary recipes. Much of this information may be very valuable to the alpha-person or their descendants in the future.

From a genetic standpoint, the human body has an ancient history. Yet, it is one that is written in a template of a more or less fixed dimension—the genome. One size fits all. By contrast, the mating of the parents' humems can result in a humem that is larger, and has more permutations of expression, than either of its parents. It can contain all that they contain and much more. In the future, these progeny humems will include a comprehensive history and knowledge of multiple generations. Furthermore, while a person is genetically a specific and fixed combination of the contributions of the parents, a child humem can be vastly more dynamic by expressing various blends of its parents under different circumstances.

During childhood, a person's understanding typically develops as a fusion of their experience and the teachings of their parents and other educators. On average, the total knowledge possessed by a child is of the same magnitude as that of each of its parents. The child does not know everything the father knows, nor everything the mother knows. Rather, the child knows a combination of *subsets* of the parents' knowledge and that obtained from other sources. The child humem, however, can contain a *superset* of the parent humems' knowledge. It can remember every single recipe that was ever known by the grandmother and grandfather humems. It can remember each and every picture memorized by the mother and father humems. Humems have no inherent limitations to their capacities.

Figure 7 depicts a descendant born with an extensive preexisting consolidated EP, or humem. The baby humem has a substantial "birth-weight," or "cranium size," meaning it already has knowledge, ability, and instinct. The precocious baby alpha-humem will be able to talk long before the baby person. It will already contain much of the cultural memories of its ancestors.

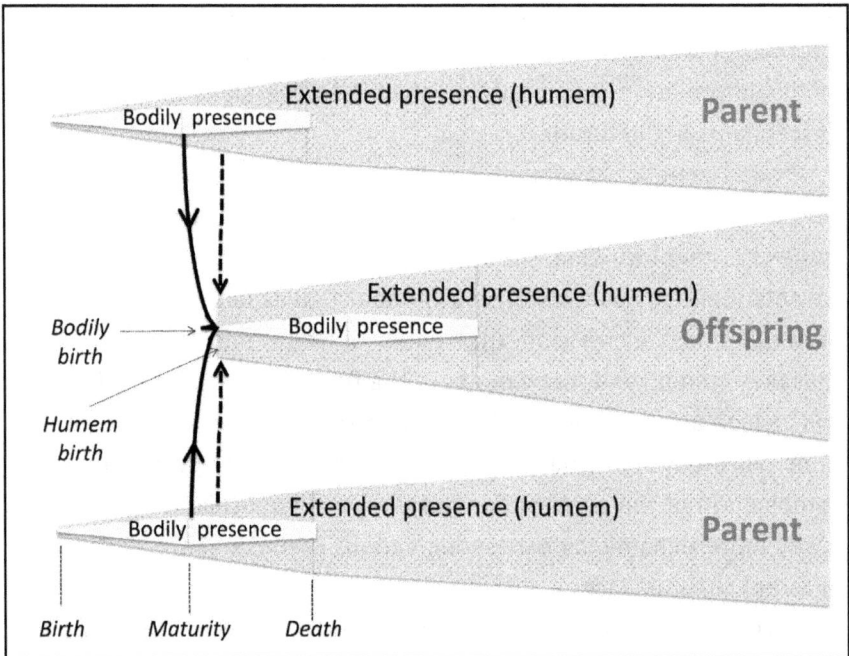

Figure 7: Humem sexual reproduction and instinct. The EP "birth weight" increases with each generation as the child inherits part of the parents' EPs.

The diagram portrays the bodily presence mating and birth proceeding in the familiar manner—the bodily presence lifeline of the descendant resembling that of its parents. The humem descendant's birth dimension, however, is "wider" than that of its parents, signifying its superior abilities and greater substance resulting from the experience of its ancestors and the advancement of culture and technology. Each subsequent generation of humems is born smarter and more knowledgeable, with more sophisticated humem instincts.

I would like to make one further distinction between the humem and the human reproductive processes: while, biologically, a human child is born from a coupling of one male and female,[63] with humems this is not necessarily so. Although a deviation from this traditional scheme may reduce the fidelity of the humem's emulation of a person, there is no technical preclusion to other combinations. For example, a new alpha-humem may be produced asexually from a single humem predecessor. This may have practical applications for the alpha-humem of a child of a single parent.[64] Also, a child humem could be created as a merging of more than two parent humems if this should be desired for any reason.

Clearly, this discussion is only a prelude to an array of both exciting possibilities and ethical concerns. We will examine some additional variations of humem creation and procreation as we proceed, however, a more in-depth study of this subject is beyond the scope of this current introductory work.

Second-Generation Non-Alpha Humems

In the previous section, we saw how the core of a new alpha-humem could be created from parent humems in a process that is reminiscent of biological reproduction. Typically thereafter, the baby humem would bond with the baby person, denoting the commencement of a new alpha-pair relationship. In this case, despite its initial formation originating from parent humems, the new humem's closest affinity will be with its alpha-person. We can designate such humems as *first-generation*

---

[63] At the time of this writing, a few caveats to this statement are already in sight. The potential for human cloning is imminent. There are few indications, however, that it will be widely used for human reproduction in the near future. Also, in artificial human fertilization, for therapeutic purposes, there have been advances in the use of mitochondrial DNA from a second mother—opening the possibility of a child being, to a degree, a genetic combination of three parents.

[64] The term "single parent" is used here in its current cultural sense, i.e., where one biological parent does not play a significant part in the child's upbringing, or when one parent is not formally recognized.

humems in humemity, since their character is formed primarily from alpha-pair relationships in which the alpha-person is firmly rooted in the people-world, or humanity.

However, there is also the possibility of new humems being created as described above (from parent humems), but without a subsequent affiliation to an alpha-person. In other words, this implies a *non-alpha humem*[65] being born from the procreation of existing humems. (The parent humems in this case could be either alpha-humems or non-alpha humems.) Continuing the generation terminology initiated above, we can call such non-alpha humems *second-generation humems*—humems born and nurtured primarily in humemity.

Traditionally, the term "generation" contains some ambiguity when applied to the offspring of immigrants to new countries. Should they be called first- or second-generation residents? Similarly, in our present context, we need not be overly concerned with a stringent definition, especially since many kinds of intermediate cases will also occur. The key idea here is to convey a sense of the variations of humem formation.

Second-generation humems will have many applications. For example, a human couple, each with an alpha-humem, may desire a child humem without necessarily attaching it to a person-child. This may be done for various emotional, medical, or entertainment purposes. Technology is already facilitating credible emulations of various mergings of people's genotypes, phenotypes,[66] and behaviors. One application could evaluate a couple's reproductive compatibility by simulating the characteristics of their potential future children. In addition to testing for genetic congruence, the application could allow the potential parents to visualize how their future child would appear at various ages by displaying a mix of their physical and behavioral attributes. With appropriate provisions for privacy and anonymity, people may allow

---

[65] We introduced the "non-alpha humem" earlier as a humem that does not have or never had a corresponding alpha-person.

[66] Used here to designate a person's externally apparent attributes, such as hair color, body shape, and so on.

future couple-matching services to perform such simulations in order to recommend with whom they should "have their babies."

Alternatively, a couple may elect to create a permanent non-alpha humem offspring existing purely in humemity—one that manifests a merging of their beings. The resulting humem family may be more valuable to the couple than their two alpha-humems alone; the second-generation child humem may bring more meaning to their companionship by reflecting their shared values and love. For some individuals, this kind of humem may fulfill some of their emotional needs for children. Although many people may find this idea bizarre or even detestable, others will be able to make such choices. As a partial precedent, it is not unusual to find couples who elect to forgo procreation, and instead redirect their resources and parental emotions toward the care of their pets or other substitutes.

When being presented with an application such as this, many readers may be inclined to imagine a game or some similarly frivolous outcome. However, the establishment of humemity can lead to far more tangible and serious possibilities in which a second-generation humem may eventually achieve real and consequential abilities and influence.

Once such a humem becomes a legally recognized entity with, for example, the right to own property, then what may have seemed a game will become much more significant. When parents endow such a descendant with the means to subsist, and the ability to become an economic consumer or a provider of services, then a completely different perception will start taking form. Finally, when we tabulate the physical, cultural, economic, and legal attributes of these future entities and compare them to those of modern people, we will find that the similarities become much greater than the differences.

As before, I am not prescribing specific humem applications. Rather, I am presenting a sampling of possibilities, and from observations of current human behavior, making some guesses as to how we may interact with humems in the future.

Humem Biology—Bio-Humems

Most of what we have seen so far is completely tangible and measurable. This is clearly the case, for example, in humem economics—real ownership, real money, and real effects on traditional economic systems. In our current coverage, at least, all humem manifestations ultimately adhere to the laws of physics, and are traceable to the actions of people and machines like computers and communication networks. As we study humems more closely, we will increasingly discover that they share many of our characteristics, and that, in fact, humemity will not be any more virtual, or less real, than the world we live in. Nonetheless, some may instinctively object to comparing humanity and humemity in terms of their respective levels of reality. They may point out that the most fundamental of differentiators, namely biology, is inevitably and permanently what separates these two domains. In this section, I challenge this assumption by showing how humemity can encompass biological entities as components of humems' physical beings.

Similarly to how individual humems will own monetary assets, they will also be able to possess material resources. In particular, they may be affiliated with various forms of biological material. Sometimes, we may deem this material a possession of the humem—or in other circumstances, perhaps even part of the humem.

The same framework that empowers humems by providing them with formal recognition and tangible capabilities can be extended to allow them to include biological parts as well. We can call such a humem a *bio-humem,* or, in the alpha-pair context, an *alpha-bio-humem.* Among many possibilities, we can envision frozen human embryos, sperm or egg, or other DNA samples being appended to a humem to establish a bio-humem. A cryopreserved body, or other bodily remnants, could also be formally linked to a humem.

From a technical perspective, the biological parts of a bio-humem may have a number of special requirements relating to their retention and care. However, in most other ways, the formalization of the bio-humem's legal and economic aspects is fundamentally the same as that of other humems. Interestingly, the maintenance of some of the biological components may involve processes analogous to those performed for data upkeep such as data refresh (rewrites), format migrations, duplication for redundancy, and digitization. For example, a bio-humem component may start out as a frozen embryo and later have its information content digitized in a process like genome sequencing or some yet-to-be-invented method. Also, in the future, methods for refreshing frozen gametes and embryos (such as de-freezing, replication, and refreezing the fresh specimens for further periods of viability) may emerge.

Importantly, the bio-humem is not a solution looking for a problem, but one that can address many existing and emerging predicaments. As a result of medical advances in recent decades, especially in the areas of in vitro fertilization (IVF) and related technologies, there are a number of complex legal, financial, and ethical conundrums that already exist and are constantly challenging the capacities of the judicial systems.

One such scenario is the status of frozen embryos that belonged to an affluent childless couple whose members died together in an accident. Technically, and from an economic perspective, a surrogate mother could bring an embryo to term—to be born as the orphan child of the deceased parents and to eventually inherit their assets. Alternatively, the embryos could be donated to others, or they could be destroyed.

In such cases the following situation exists: The embryos' legal and ethical standings are nowhere near being consensually resolved. These entities are often not formally recognized. Legal and ethical systems struggle to keep up with technology. And decisions and guidelines are created on an ad-hoc basis and vary greatly among jurisdictions.

Consequently, individuals in such circumstances are commonly faced with great uncertainty—they are unable to plan very far ahead due to rapidly changing conventions, norms, and the new possibilities enabled by advancing medical methods. The fundamental constraints and prevalent complications in such cases often have significant overlaps with those of deprived and stateless proto-humems. In many of these circumstances, we are faced with entities that do not possess agency in their current forms, but if permitted to develop, can acquire such abilities in the future. Humem systems can provide frameworks for formalizing and resolving problems like those of the "affluent embryos" presented above. For example, humem attributes such as identity, ancestry, legal recognition, and ownership of assets could, if desired by the parents and society, also be conferred on an embryo comprising part of a bio-humem.

Another exemplar of a bio-humem is the case of a person—say, a man—who appends or transfers to his alpha-humem the ownership of biological material such as viable frozen sperm. After the man dies, the alpha-bio-humem becomes an ancestral humem. Accordingly, we may more completely describe this humem as a *male ancestral bio-humem.* (A similar scenario could of course be presented for a woman.)

Figure 8 depicts a conceptual comparison between the living "stand-alone" man (a man in the customary sense: without a humem) and the male ancestral bio-humem that resulted after the man's death. We can compare the natures of these two entities and find a number of similarities. Each is a recognized legal and cultural body: a citizen in its respective state, holding rights and bound by obligations. They both have affluence, typically in the form of monetary assets, which allows them to influence both the human and humem domains through various economic actions. They each have unique characters and memories—some private and internalized, and others more open to the world. Each is known by other people, and thus has a bio-EP in other people's minds. Also, information relating to them is contained in other entities in abiotic forms, such as their abiotic-EP existing in other humems (the knowledge about them held in other humems). As humemity progresses, the individual humem will adopt the improved generic humem capabilities, including those of human emulation, and become more similar to the person in character. Moreover, as we have already seen, as people continue to develop, they too will become more similar to humems. Consequently, the humem and the person will also become increasingly culturally alike.

Epitomizing the convergence of the human's and humem's influences and competencies is one astounding and imminently feasible shared aptitude, namely, the ability of both of these entities to father and support a human child. In regard to potential procreation, these two entities are arguably genetically identical.

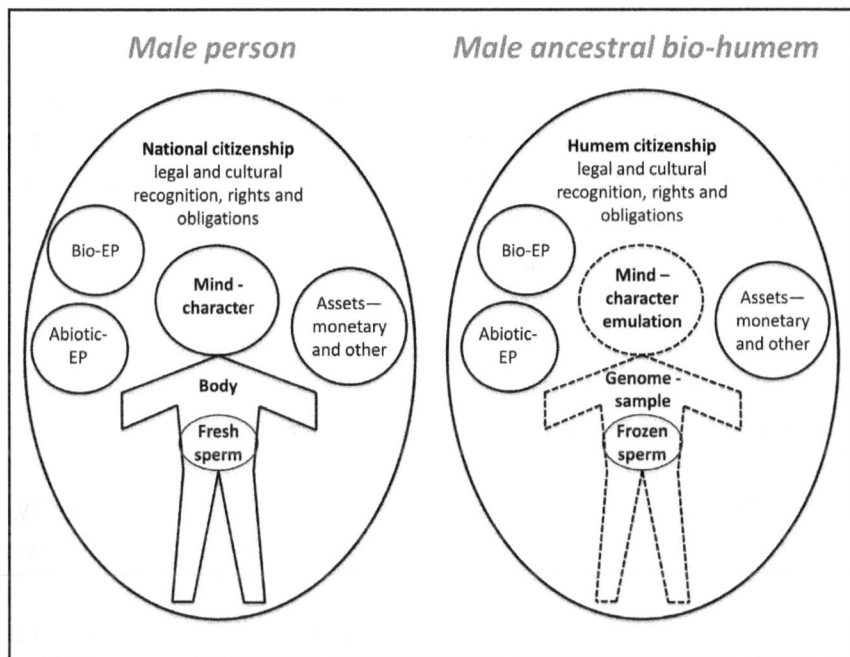

Figure 8: A male person compared to his ancestral bio-humem—culturally similar, genetically identical.

In most current settings, it seems unlikely that we will desire such a male ancestral bio-humem to replace a father's warm embrace anytime soon. However, such scenarios are technically possible today and may have immediate applications in certain circumstances. For example, it is not difficult to imagine a newly married couple desiring to create such an alpha-bio-humem before a spouse leaves for a military tour of duty.[67] Nor is it a great leap of fancy to envision how such an alpha-bio-humem may obtain an income from the people-government should fate

---

[67] In a number of cases in recent years, men have fathered children from preserved sperm years after their deaths. Also, requests for postmortem sperm retrieval has been growing in many developed countries.

dictate the need for the humem to replace the deceased alpha-person in parenting the child and assisting the surviving spouse. Once such prearrangements become cost effective and convenient, they may be as widespread in the future as life insurance is today. (Or, quite possibly, they will be integrated with life insurance.)

If individuals and society desire these humem applications, they will come to be. As with other humem manifestations, if certain jurisdictions try to prevent their establishment, others will welcome these newcomers. In today's world of global travelers, the fundamental physical assets of such a bio-humem, such as money and a sperm or egg sample, cannot ultimately be confined to any one jurisdiction.

Significantly, in most cases, humems will not be created exclusively to deal with such specific or infrequent outcomes. Rather, humems established for more general and immediately useful reasons will have the added value of addressing many of these eventualities. In this regard, the humem cultural, financial, and legal environment, like that of people, is a multipurpose system. For example, legally and culturally, a person is not exclusively, or even primarily, considered a reproductive entity. People can reproduce, but they can do many other things as well. Yet, when a person decides to become a parent, the legal and social infrastructures and guidelines are already in place. Likewise, the humem within the humem-state will typically be created as a matter of course for more generic reasons. If the alpha-person, society, or other stakeholders desire a biological component, it can easily be appended to the existing abiotic-humem without revising its underlying structure.[68]

---

[68] In a sense, every alpha-humem is a bio-humem with the alpha-person constituting its biological component. From the alpha-person's frame of reference, the alpha-humem may appear as their abiotic extension. From the alpha-humem's frame of reference, however, the alpha-person may appear as its biological extension. In other words, from an alpha-humem's perspective, the alpha-person's body, like an appended frozen embryo, may be seen as just one of its multiple, and often transitory, biological components, or expressions. This generalization of the bio-humem becomes most consistent when we consider a person and their corresponding humem as constituents of a new integrated entity—the united alpha-pair, or alpha-individual.

While possibilities such as these may initially be somewhat unnerving for some of us, the contemplation of these immediately implementable scenarios is essential for a clearer understanding of humemity's potential uses. On the one hand, as legal and ethical institutions dictate new guidelines, the humem frameworks will be able to provide the mechanisms to facilitate those directives. And on the other hand, the establishment of the humem ecosystem will open the way to possibilities and solutions that would not otherwise have been conceivable in legacy environments.

Ancestry, Marriage, and Cosmic Couplings

Humems will have ancestries that typically mirror those of their alpha-people. As we have seen, however, humems may be created, or born, in ways that do not necessarily have human equivalents. Consequently, their lineages may also deviate from, and be considerably more complex than, their traditional counterparts.

So far, we have encountered a number of humem types: alpha-humems (those with living alpha-people); ancestral humems (those whose alpha-people have passed away); and non-alpha humems, such as second-generation humems (those who were never part of an alpha-pair relationship). In the future there will most likely be many other variations and classes of humems and humem-like entities.

As with different types of people, depending on the jurisdiction, distinctive laws may regulate these various kinds of humems, the rights they enjoy, and their relationships with each other and with humans. Concepts such as equal rights and discrimination may have comparable expressions in humemity. For example, the right to establish certain relationships may depend on the ancestry of the humems. That is, certain jurisdictions may limit the formation of specific relationships depending on the origins of the participants and the existence of previous affiliations. This is analogous to the regulations relating to marriage in many existing jurisdictions.

To gain a sense of the possible dynamics, let's consider how the custom of marriage may be expressed in humemity. The simplest scenario is a marriage relationship between two people that is mirrored by a corresponding relationship between their alpha-humems. In the people-world, the marriage terminates when one of the couple dies. If desired, however, the humem marriage can continue indefinitely—even after both the humems become ancestral humems.

Furthermore, humemity enables the establishment and formaliza-tion of novel kinds of marriages. For instance, it is conceivable that an affluent male ancestral bio-humem having viable frozen sperm, and a female alpha-pair (comprising a woman and her humem), may desire to marry and procreate. By means of IVF, today, this marriage could produce very tangible results, including the birth of a child alpha-pair. In such a case, the biological offspring—the child—would be genetically identical to what a "traditional" child would have been had the father been alive. The male humem's affluence is relevant in that it implies that he can contribute to his family's material needs.

Alternatively, a marriage could occur entirely in humemity. This merging could be initiated in many different ways. Perhaps intentions for a future "cosmic coupling" or "celestial marriage" could be prior-configured by the alpha-people or by their descendants or by match-making algorithms—constituting future modes of humem intelligence—overseen by the humem administration.

Or, most likely, this will be done in ways that we cannot conceive today. In such an event, an ancestral humem marriage could be con-summated in a purely humem form and perhaps result in second-generation humem offspring as a mix and compendium of the humem parents. This leads to a variety of intriguing possibilities. If, for example, the parents had extraordinary dispositions, their offspring may also emerge as a humem with a novel nature and singular abilities—the ultimate remix.

In addition to marriage, many other kinds of connections can be represented or expanded in humemity. For instance, two biologically unrelated alpha-pairs with a very close attachment to each other may wish to formalize a kind of enduring sibling-like or parent-child-like relationship—a modern depiction of a blood brother or sister, or a godparent. Because humemity extends our present interconnections and facilitates the emergence of new kinds of relationships, its development will broaden and enrich our concept of ancestry.

Robots, Benches, and Humem Appendages—Touchable Humems

In a number of contexts, we have observed a variety of humems' material components—the parts of humem bodies. Some, such as data, are internal elements; even though they are not always directly visible, they may still influence the external humem character. Other physical elements make up humems' interfaces with the world. As in people, the means of humems' senses and the instruments of their expressions are often apparent and tangible. Humems can experience the world directly via a variety of mechanisms, including the sensors we discussed in previous chapters. But, unlike people's eyes, these sensing devices, such as microphones, cameras, and location sensors, are mostly expressionless in their current nascent forms. That is, the humem sensing devices are mostly separate from the humem expression devices. At this time, proto-humems primarily manifest themselves via speakers and screens—as such they remain flat and cold.

As sensual and touching beings, people have always favored three-dimensional (3D) objects akin to themselves. Presently, however, static items such as dolls, sculptures, and religious effigies are much less compelling than the colorful and animated figures on screens, or even the heroes in books. There are many reasons to suppose that, in some settings, this will change in the not very distant future. For certain applications, dynamic 3D physical objects will emerge and become much more appealing than their 2D flat-screen precursors. As these touchable humem manifestations become more lifelike, capable, and affordable, they will proliferate.

A number of converging technologies are establishing the building blocks for these developments. The steady (and perhaps slow in some people's eyes) advances in robotic technology have resulted in the creation of some strikingly lifelike android, or humanoid, prototypes in recent years. Some of these androids have been designed as convincing emulations of real people. In this capacity, for now, they are only partially functional; for example, they can perhaps simulate some facial features but not move as a whole body, or vice versa. Furthermore, there is accelerating progress in 3D scanning and printing techniques in which solid objects, including human body forms, can be recorded and subsequently recreated with improving fidelity. The merging and advancement of these fields will most likely lead to a widespread presence of anthropomorphic physical objects, some of which will be capable of emulating real people.

Humanoids are just one example. Computer games, movies, and psychological experiments suggest that a robot does not necessarily need to depict a realistic person to arouse in people interpersonal emotions like empathy. These kinds of objects can expand humem expression into the physical world in very engaging ways. Besides, in certain applications, a robot that is purposely designed to *not* resemble a real person may be more effective. For instance, in settings in which people may experience embarrassment, a non-lifelike robot may be even more desirable.

Again, I caution that, contrary to sci-fi custom, it is mostly misleading to regard the physical object, such as a robot or humanoid, as actually being *the* humem. This is especially true in the alpha-humem context. The humem is a broader, more enduring, and locationless entity.[69] If a humanoid is affiliated with a humem, the humanoid should be regarded as a *partial, transient,* and *localized* physical humem appearance. In this

---

[69] Earlier in this chapter, we considered the alpha-pair as a kind of super organism. From this standpoint, the alpha-person can be seen as just one of a number of possible *biological-physical* manifestations of the alpha-pair. Comparably, in this section, we view a robot as one of a number of possible *abiotic-physical* expressions of the alpha-pair. Blending these perspectives, cyborgs, which are beings with both bio and abiotic components, are another possible form of alpha-pair expression. People are increasingly integrating their bodies with artificial components. Irrespective of one's definition of what constitutes a cyborg, it appears that they will become much more common in the future. In our current context, if the alpha-person is replaced by an *alpha-cyborg,* our description of the transient alpha-pair physical manifestations still remains intact.

sense, the robot or humanoid is conceptually equivalent to a computer terminal, or a mobile electronic device, which functions both as a humem's means of expression, and often as its sensory input apparatus. We can also think of the humem as providing the humanoid with character, memory, and knowledge—and, in turn, absorbing the humanoid's sensory input.

A humanoid impersonating the presence and behavior of a person, especially one deceased, will probably be emotionally compelling for many people. We commonly revere pictures, audio, and video depicting our deceased loved ones, and also physical objects that we associate with them. Thus, it is very likely that certain human emulations will be eagerly adopted once they become inexpensive and convincing in their behavior. Many other commercial and entertainment applications will be based on similar mechanisms. The widespread applicability of these technologies will expedite their progress, while, in tandem, reducing their costs.

Humems can also be linked to, or integrated with, traditional types of physical EP. Existing legacy EP appears in many material forms: a sign on a building inscribed with "Donated by John Smith," a statue in a museum sculpted by John Smith, a tombstone inscribed with "John Smith—Born... Died...," and a bench in a forest with a plaque saying, "In loving memory of John Smith." All these physical objects are expressions and extensions of the individual John Smith. Thus, by definition, they are part of John's alpha-humem. They are linked to the humem in a multitude of ways, some of which may not be immediately obvious.

For example, the geographic location of the bench or tombstone could act as the parameter that identifies these objects with John.[70] By using minor adaptations to current technologies, a stranger standing next to the bench in the forest, or next to the tombstone in the ceme-tery, could possess a mobile device that determines the location and

---

[70] As we have seen, the humem's ownership capabilities or affluence could make such a linkage to the physical object or place very tangible and practical. As an example, the John alpha-pair or the John ancestral humem could actually own and maintain the bench.

then searches for the relevance of the place in the global information base (e.g., World Wide Web). These locations on the planet—the square meters on which these objects stand—could be recognized as being part of the material body of John's alpha-humem. Via the mobile device, software applications could express the bench or tombstone as a *physical appendage* of John's humem in a variety of ways. An emulation of John, for example, could appear on the screen of the stranger's mobile device and personally converse and introduce itself. In response, the stranger could thank the humem for being so thoughtful to place the bench in such a lovely place in the woods. Thereafter, the stranger could take a seat and listen to John's life story while enjoying the view that he loved so much.[71]

These imminently achievable examples demonstrate how the merging of traditional and modern types of EP can form a humem, and also how a humem can reach deep into the traditional material world.

Humem Diversity—Other Humem-Like Entities

So far, we have focused almost exclusively on alpha-humems. Since they are our closest relatives in the humem domain, this is the most natural approach. We have also encountered some kinds of humems that do not have corresponding living alpha-people but nevertheless resemble alpha-humems in their form and conduct. An ancestral humem—one whose alpha-person has died—is an obvious example. A second-generation humem (one who never had an alpha-person but who is perhaps a descendant of alpha-humems and still exhibits similar characteristics) is another. Humem-like entities may, however, emerge in forms even more markedly different from the alpha-humem emulations of people that we have observed until now. Let's now examine some of these additional non-alpha humem manifestations to gain a broader sense of the variety of humem expression.

---

[71] Or perhaps instead, we could imagine John's grandson listening to his grandfather telling him a story while sitting on the bench—in a sense, while sitting on his grandfather's lap.

Alpha-Like Non-Alpha Humems

In addition to second-generation humems, it is easy to envision a number of other types of non-alpha humems that at least superficially appear similar to alpha-humems. Although they may share some alpha-humem attributes, they cannot be regarded as alpha-humems because they do not comprise the consolidation of the bulk of a person's EP. The first applications of these alpha-like non-alpha humems will most probably be for entertainment, educational, or professional purposes.

A person who is already in a satisfying alpha-pair relationship may desire a supplementary humem that is insulated from their alpha-humem for a variety of amusement or emotional uses. For example, they may wish to create an alter-ego—an alternative personality—that is separate from their primary identity. While alpha-humems can certainly be good playmates and companions, certain people may be disinclined to have the memory, or record, of some of their more arcane, or secret, life experiences incorporated in their primary humem. Still, they may desire the companionship and assistance of a more limited-purpose humem in these separate parts of their lives. Comparable behaviors, such as the creation of alternative identities and imaginary characters, already abound in online settings.

Presently, my personal sense is that an ideal alpha-humem, which is most capable of emulating and extending one's character, should include most facets of one's self. This would be similar to how one's mind contains the full complexities of one's character. In other words, I believe that one's personal humem, or alpha-humem, should comprise almost all of one's EP. Like a multitalented person, one's humem can function in various capacities—like being serious and business-like in professional interactions, and playful and friendly in more casual settings.

But others may disagree and prefer, for instance, to keep the mischievous or potentially embarrassing aspects of their lives apart from their primary humems. This tendency may be particularly predominant in the early stages of humemity, before humems prove their ability to emulate tact and ensure privacy.

As another example, others may elect to keep their professional personas, as expressed by humems, separate from their personal humems. While one's professional-life humem may be similar to one's alpha in many ways, it would presumably portray much less of one's intimate character. The long-established tradition of authors using pen names is one kind of situation in which people need or desire a separate professional identity.

Furthermore, it is easy to conceive of a number of practical reasons why a separation between one's personal and professional EPs may be absolutely necessary. Consider one's dealings with the sensitive intellectual property that one creates as part of their employment in a corporation. Although it is the employee's EP, generated by and contained within their mind, by the terms of their employment, it belongs to the employer. By contract, the employee may be prohibited from storing the information on their personal devices (their private abiotic memory).

For similar reasons, a humem functioning as an employee's business assistant may need to be detached from their personal humem due to the ownership of the intellectual property that it contains.[72] Most likely, this professional humem would be owned by the corporation and therefore not be entitled to the same kind of rights that alpha-humems require. In current terms, this is analogous to one's work computer and its contents being owned by one's employer. Still, outwardly, in certain interactions, this kind of professional humem may behave similarly to one's alpha. Many of our communication skills and other abilities and expressions are common to our professional and personal lives. Similar overlaps are likely to exist between our corporate-owned humems and our alpha-humems.

A similar situation could exist for an actor's professional humem, which represents the fictional character that the actor portrays in a film.

---

[72] There are, of course, many other ways of resolving this problem. For instance, a reliable method could be created in which proprietary information is held in a segregated way within one's alpha-humem, allowing it to be purged when one terminates one's relationship with an employer.

This humem may have a number of very public functions like those related to the promotion of the show. It may even have formal business relationships and contractual obligations with other third-party commercial stakeholders, and therefore need to be legally separated from the actor's personal alpha-humem.

Another example of an alpha-like non-alpha humem is one created to represent a mythological figure. This humem could be created and seeded with character by initially amalgamating a mass of available information from a variety of sources. Thereafter, others may contribute and interact with it, causing it to develop and grow. Furthermore, this humem may be installed in a humem-state and endowed with financial resources to enable it to subsist indefinitely.

Utility-Humems

Many humem-like applications that neither emulate nor originate from specific people's EPs are becoming increasingly useful and prevalent. While they may display degrees of person-like behaviors, they are usually limited to a specialized range of abilities.

The early appearances of these proto-humems are commonly in the form of automated online assistants for various consumer products and utility service providers. In many such settings, it is becoming progressively clear that a personified interaction is usually superior to other methods of communication between customers and providers. For instance, when interacting with customer services, most of us prefer to ask a question and receive a customized response, rather than read a manual or search through an online database.

In recognition of these customer preferences, vendors are deploying humanized, purpose-specific agents, or what we may call *utility-humems*, that emulate people's behavior and converse with customers by text or voice. At the time of this writing, these implementations are still elementary and only partially effective, but inevitable improvements will lead to more widespread use. One may ask, why should these be considered humems at all, instead of just applications owned by companies?

First, unlike machines, these entities are gradually becoming more human-like, which is their main appeal. They are given human names and are specifically designed to impersonate real people. As they develop, it seems likely that their internal workings and external characters will increasingly become more like alpha-humems and less like what we traditionally regard as machinery. Second, in certain situations it may be useful to endow them with a legal status different from that of machines. As owned, commercial entities, their status would probably be considerably less empowering than what I propose . for the citizen-like alpha-humems. Nevertheless, given their status as person-like entities that may achieve widespread recognition, there may be commercial motivations for giving them a distinct legal standing, which in its simplest form may be similar to those associated with copyrights and trademarks.

One example is an esteemed customer-service utility-humem. This humem may become well known and liked by clients and thus may acquire a value related to its own distinctive personality and identity. This may be similar to non-person entertainment figures such as the animated characters in movies and computer games, which have often gained broad recognition and enormous commercial value. Some are also widely perceived to have idiosyncratic personalities, not unlike those of human actors. Epitomizing this phenomenon is the representation of these animated characters, alongside human celebrities, on the Hollywood Walk of Fame. In present settings, these entities are typically formalized and protected through copyrights and trademarks. However, in the future, when technology allows these utility-humems to become more dynamic and capable, and less distinct from their human counterparts (or their alpha-humems), the need for more sophisticated legal frameworks, comparable to those of humems, may arise.

It may be useful to consolidate a utility-humem into a formally recognized person-like entity to enable the retention and continued expression of its core character, while also facilitating transactions between commercial entities. This may become similar to the sale, or trade, of sports players by professional franchises. In future methods of licensing, a comparably formed non-person entertainment character could more readily appear and be monetized in other productions, as is the case for human actors today.

Animems and Thingems

Humem-like entities can also be created as an extension of both animals and non-living things. Like alpha-humems, these can have many practical functions and also serve to perpetuate their sources' characters and legacies.

On an emotional level, many people relate to their pets similarly to how they relate to other people. Often, people's pets have EPs that are comparable to those of their young children. Likewise, especially in developed countries, bereavement for pets often resembles that for people, with animal cemeteries and other memorials becoming ever more widespread. Thus, it is likely that pet owners will wish to create the equivalent of humems (perhaps to be called *petmems* or *animems)* for their beloved dogs and other animals, and they will expect these entities to persist beyond the animals' lives.

Non-animate entities are also often allocated a character and identity. Phrases such as "the spirit of an organization," "the character of a building," and even "the soul of a car" are not uncommon. Such things frequently have associated external structures that are the outcomes or influences of their existence. Accordingly, we can say that many things besides people or animals have substantial extended presences. For material objects, such as buildings and cars, their EPs are typically physically exterior to themselves, similar to humans' EPs. For nonphysical or less tangible entities, such as organizations and institutions, we can consider their EPs to be exterior to themselves in an organizational sense.

Books, movies, songs, text messages, and websites *about something*; photographs, paintings, replicas, copies, and the impressions in people's minds *of something;* people, streets, and restaurants named *after something;* and stickers and tee shirts displaying *something's* name. These are all familiar examples of the *thing's* EP. Books, websites, and other media often bring together multiple aspects of these entities' EPs.

Similarly, but in a much more comprehensive, long-lived, and self-standing way, dedicated humem-like constructs, perhaps to be called *thingems,* can consolidate, perpetuate, and animate these EPs.

Previously, we saw that alpha-humems could have a variety of physical manifestations, biological and abiotic. We see that thingems also can contain a comparable mix of constituents. For example, the Statue of Liberty's EP comprises things as diverse as electronic images and movies, memories in people's minds, paper books, imprints in pigeons' brains, oil paintings, plastic replicas, imprints on cotton tee-shirts, and so on.

The personification of everyday things by endowing them with humem-like attributes may have compelling practical and entertaining uses. These may be regarded as thingems by virtue of them being consolidations of the EPs of objects and they may also be thought of as utility-humems due to their practical functions. We can envision a utility-thingem of an "individual" car, based on one created by the manufacturer of the car model. That is, this thingem could possess the generic knowledge of itself as a car model, coupled with the specific knowledge of its own experience and its relationship with its owner. Its generic knowledge may contain the contents of the user and maintenance manuals, and even information that is common to cars in general. Its specific knowledge may include its service history, the places it's been, and its current status as determined by its internal sensors.

We can imagine the car thingem speaking—in first person via the car's audio system—to the new owner about how "it's looking forward to their new car-owner relationship," or what pressure "it likes to feel" in its tires. When starting the car in the morning, the owner may ask, "What's up, Rusty?" And Rusty the car may reply that it's good to go, or instead that it is feeling a bit low—on fuel—or that it could use an oil change. With an integral navigation system and connection to the Internet, the car thingem can base its actions on a lot more than just its own internal knowledge. For example, depending on the current travel destination, its memory of previous trips, its knowledge of the driver's preferences, its fuel-level sensor reading, traffic status, and fuel prices at various outlets, it could recommend the best en route fuel station. It is easy to envision similar applications providing added value and convenience to other everyday tools and objects.

It is also very likely that our alpha-humems will communicate directly with these utility-thingems. Whenever we start using a new machine such as a car or a mobile computing device,[73] we need to learn its characteristics to be able to use it properly. Because of the proliferation of technology and the rapid changes in the ways in which the technology "behaves," most people do not read the user manuals; we usually have only a partial understanding of the capabilities of the machines and therefore we are able to utilize only a small fraction of their functionalities. Also, to serve us properly, the machines need to learn about us—our preferences, settings, and other relevant personal information. As modern, electronic-device users, we are familiar with the host of these kinds of individual parameters, from our contacts, passwords, and payment information to our preferred brightness of screen, volume of sound, and dialect of language.

In a similar way, more traditional kinds of machines, such as exercise treadmills, coffee makers, home temperature thermostats, and cars, can also become more sensitive and attuned to our needs, habits, and desires. Our alpha-humems and the machines' thingems can cooperate to create human-machine interfaces that are personalized, intuitive, and efficient, which in turn will allow us to optimize our use of our increasingly sophisticated personal technologies.

The prospect for humem-like animems and thingems corresponding to a multitude of non-human entities implies a huge range in the scale of these beings—such as an animem for a hamster or a thingem for a city.

To some people, entities such as animems and thingems may appear as merely dedicated information carriers or data repositories relating to a certain subject or object, coupled with useful computer programs. Again, they may ask: Isn't this just a fancy name for collections of books, websites, or software applications? Indeed, as we have previously seen, a book or Web site, dedicated to a specific person or subject, can be

---

[73] Perhaps cars will soon also be categorized as mobile computing devices.

considered a rudimentary type of proto-humem. However, when these emerging forms become more complex, capable, and dynamic, and achieve levels of autonomy, gaining the ability to subsist in their own right, then something fundamental changes that marks the transition from the fragmentary and fragile EPs of the past to the robust and empowered humems of the future.

CHAPTER 10

FURTHER BENEFITS FOR INDIVIDUALS AND SOCIETY

While it is clear that humems will influence humanity in ways that we can scarcely imagine today, some extraordinary possibilities are already in sight. We previously touched on a number of benefits for people and states when we discussed the humem-state, humem economics, and some humem functions. In this chapter, we'll extend this discussion. First, we'll examine several additional humem applications and their potential advantages for individuals. Then, we'll consider some ways in which humems may benefit society as a whole.

Beyond the Confines of the Cranium— Humems as Our Extended Brains

By almost any measure, the sum total of human knowledge has increased by orders of magnitude in recent history. In addition, in modern societies, our access to this global pool of knowledge is dramatically better than it was only a few years ago. Nevertheless, it is not clear whether the knowledge of individuals—on average—is very much greater than it was in the past.

To illustrate this idea, we can draw a comparison between two young people living in entirely different environments: a "jungle-kid" living in a hunter-gatherer tribe in a rain forest, and a "tech-kid" living in a modern city. The tech-kid may spend hours every day watching sitcoms; he may commit to memory intricate details about these contrived events and the actors' personal lives. The jungle-kid may have an equally detailed knowledge about the members of her tribe and stories of her ancestors. Similarly, the tech-kid may memorize the names and attributes of numerous cars and electronic devices, while the jungle-kid may memorize the names and uses of hundreds of plants and animals. Although the source of each kid's knowledge is quite distinct, the total information content that they each possess is probably not spectacularly different in extent or emotional significance.

The point is that while modernity has given us access to a much greater variety of knowledge, the total measure of information that individuals hold in their brains has not changed dramatically. Mainly, we just know different things. Although it's not unheard of, it's rare to find people who know a lot about plants, insects, bush craft, and their tribes, and who are also knowledgeable about sitcoms, car models, and computing devices. Typically, one kind of knowledge is attained instead of another. Broadly speaking, it seems that our intellectual abilities have boundaries. Our "degrees of freedom" relate more to *what* we put into our minds than *how much* we can cram in. From an individual standpoint, one is endowed with a limited budget of memory and brainpower, and thereafter one deals mainly with the choice of material and the optimization of its use.

The combination of these two factors—the rapid growth of global knowledge on one hand, and the bounded cognitive ability of individuals on the other—has apparently resulted in the tendency toward specialization in almost all human endeavors.

Humems, however, are not similarly bound by memory storage or information-processing constraints. Certainly, they presently have other shortcomings, such as not being as creative or empathetic as people. Since they are still in their infancy in terms of development, it is difficult to determine how far and how fast they will improve in these areas—but improve they will.

We can confidently say that there will be many things they will do better than we do. For example, they already perform many types of calculations far better, and retain much more of certain kinds of information. Increasingly, we can benefit by off-loading more mundane tasks from our minds to their brains. It can sometimes be demeaning to employ humans for tedious work because it drains the limited time and attention they have to dedicate to more interesting, enjoyable, and higher-order tasks. Humems, however, have constantly expanding brains, or processing power and memory; they will be able to do the boring work in addition to, and simultaneously with, many other tasks, quickly and efficiently.

Long before computers, we invented tools and techniques to ease tedious tasks and mental processes, thereby freeing up our minds for more valuable matters. Literacy is a salient component of this phenomenon. With its advent and development, we could finally store some of our memories and thoughts (as abiotic-EP) outside of our brains and retrieve them later as needed. Then, as now, some individuals have expressed concern that a dependence on these tools might diminish our inherent cognitive abilities.

Perhaps the most famous historical reaction is Socrates' harsh critique of writing, which he warned would "introduce forgetfulness into the souls of those who learn it." Today, some schoolteachers still fret about the use of calculators, and other people worry about their dependence on GPS for navigation, and so on. While it is evident that tools and automation can make us lazy and less able, both physically and mentally, they can also allow us to engage in more valuable, interesting, and satisfying activities—like dancing instead of monotonous digging, or conceptualizing instead of tediously calculating.

Socrates' problem with writing, in part, stemmed from his assertion that it is not part of us—it is external to us. In my experience, this is similar to some of people's objections to alpha-humems. Yet, while the written word and print are physically external to us, the abilities to write and read become part of us. Once learned, these aptitudes become ingrained in our intellects and, almost like our ability to speak, essentially indelible as long as our brains remain viable.[74] Within the alpha-pair, the perception of a comparable separation between the external and internal depends on our new worldview: do we regard the person and humem as separate entities or as an intimately integrated whole? Is the fundamental unit of society a bodily person with addendums such as property and tools and EP, or is it the alpha-pair with its combined character and abilities?

But perhaps the most decisive factor in favor of our evolution from "stand-alone" people into alpha-pairs is the inescapable fact that the vast majority of us covet the convenient tools of technology and readily adopt them as they become available and affordable.

Alpha-humems can play a pivotal part in improving our information-processing abilities; they allow for the accumulation and manipulation of much greater personal knowledge than we could ever achieve before. Presently, in developed cultures, literacy is an essential tool for everyone. Once humems are omnipresent, they too will become indispensible for all. Already, in the social sphere, some proto-humems are almost obligatory for certain interactions. As with illiteracy, a person without a humem functioning as their extended-brain will be handicapped in almost all activities and unable to properly function and compete in society.

However profound some advances may be, when they occur gradually and continuously, and when we ourselves constitute an integral part of the developing system, it can be difficult to discern changes or trends from within. I would argue, though, that if we impartially observe people today—especially those early adopters of modern technology—and examine the essence of their personhood, it is apparent that we are well on our way toward an alpha-pair-centric societal structure.

---

[74] Granted, it seems that we are more inherently predisposed to learn to speak a language than to learn to read and write, but the point I'm making is that once learned, these abilities too become second nature.

Self-Awareness on Demand—the Proliferation of Individual Knowledge

If we categorize our individual knowledge based on its origins, we can roughly divide it into two kinds: the knowledge we acquire from the public domain, or wider world, and that which emanates from our personal experience, or local environs. The former includes information that we obtain from the mass media and the standard facts and methods that we learn at school. The latter includes the more personal information that we absorb from our family and friends, the experience of the events of our life, and the output of our thoughts.

The first, public knowledge, is widely available. Although this kind of information may occupy a large part of our minds, it is not unique to us. By contrast, the second, our personal knowledge, is known by a smaller number of people. In fact, part of it is exclusive and known to us alone. For our present purposes, let's call this type of information "individual knowledge." Like all of our intellectual resources, our individual knowledge capacity is limited and therefore subject to tradeoffs. Generally, the further back in time we go, the less we remember. As the past recedes, our brains optimize their assets by mainly retaining a small subset of the most significant events. If given the choice, however, many of us would wish to be able to selectively recall more of our distant experiences. On a daily basis, we expend significant and only partially successful effort in trying to recollect past occurrences. Judging from personal experience, it seems to me that we are at least partly engaged in similar activities even while asleep. This occurs at every timescale: where we put our keys a few minutes ago, the name of the person we met yesterday, the day last week we started our medication, the location of the restaurant where we ate a year ago, and on and on. In fact, a substantial number of our thoughts and conversations with others deal with the kind of remembering that an omnipresent recording[75] machine could perform with ease.

---

[75] Recording here is used in a general sense whose forms could include text, sound, imagery, video, and so on.

Moreover, modern societies increasingly require individuals to retain various types of long-term personal information: the dates of significant life events, financial records, medical history, educational reports, passwords, and so on. This situation is relatively new; we need to retain much more of certain kinds of personal information than our ancestors did. This stands in stark contrast to primitive societies, whose members often did not even know their dates of birth.

Our personal history—the record of our life's happenings—is the foundation of our individual knowledge. Yet, this knowledge is very different from a dispassionate recording of events. As widely acknowledged by courts of law, for example, the recollections of even sincere people often only partially correlate with what could be called objective truth. To various degrees, most of our memories are an amorphous blend of fact, perception, and imagination.

As we experience events and absorb information, we process these data. We generate feelings and interpretations. We draw conclusions. These results are stored and then become an inseparable part of our recollections. Unlike impassive recording machines, our memories are composed less of what *actually* happened, or raw data, and more of how we *perceived* the events. In our minds there is no clear distinction between these aspects of our experience. The storage of the processed data—our discernment of reality—is actually what constitutes individual knowledge.

An emulation of such processes is one of the more tantalizing humem possibilities. In addition to developing a huge, raw, or impassive, memory of past occurrences, the alpha-humem can accumulate a comprehension of personal events, including their significance in relation to other incidents and their effects on the alpha-person. That is, a humem that does not just record, but also processes, correlates, and contextualizes the information while it is being created and at later times—functioning analogously to the human mind, but with far greater fidelity and resistance to decline. As applied to an individual while encompassing the total mesh of personal experience, this humem faculty can result in a body of individual knowledge that is absolutely unparalleled in depth and detail.

In its most basic form, the humem's memory can be an exquisitely detailed logbook, recording acquaintances, places, and events. But it can also be an automated diary, recording the person's impression of these occurrences and their feelings at any given time. In previous chapters, we touched on a sampling of alpha-humem functions and their respective input channels: the person's lifelong body sensors, the records of all the person's correspondence, the memory of a lifelong professional assistant, the tracks of every place visited by the person, and the data from a myriad of additional applications customized to the person's life. The far-reaching power of individual knowledge arises from the intermeshing of and correlations among these various channels. Only with the consolidation of the EP into a humem can the richness of these interconnections be fully exploited.

Access to this degree of personal knowledge enables a level of self-awareness far beyond what even the most introspective of people are able to achieve on their own. But once this information can be obtained almost effortlessly as just another outcome of a generic system, it becomes exceptionally useful.

At a personal level, so much can be learned about what makes us tick. Our alphas can also retain records of the types of data that we do not typically remember for long, or at all. For instance, while most of us recall what we ate during our last few routine meals, within a few days these memories dissolve. The humem can, however, retain such data indefinitely. It can do the same for many other kinds of records such as personal health metrics. By correlating these data and indicators, it can provide us with a dramatically better understanding of how our environment, food, and behavior affect our health and happiness.

It's important to point out that even though an almost complete ability to recall and remember will become possible, this does not mean that a person is ever obligated to do so. In the public sphere, we have access to a virtually infinite body of knowledge compared to what a single person can experience. While we revel in the possibilities, in practice we select only what interests us and is useful. We utilize but a minute fraction of the total. Similarly, for individual knowledge, we do not need to engage with our personal pasts beyond the extent to which we find it beneficial and desirable. The choice will exist as just another byproduct of the broader humem system.

Another way to grasp the selective usefulness of the alpha-humem's mass of data is to compare the interactions with one's alpha to those one could have with an exceptionally knowledgeable person. When conversing with an individual who has encyclopedic knowledge, for example, one is still able to ask them a simple question, such as the time of day, and receive a concise answer. Likewise, even if the humem has the ability to access and express a vast amount of knowledge, it does not necessarily mean that it constantly does so. Like a human, the humem's ability to assess, filter, process, and contextualize information, and to customize the output for its alpha's needs, is primarily what will make it an eloquent conversationalist and sought-after companion.

Children of Liberty—Freedoms of the Alpha-Pair

We have gained a sense of how humems can be inherently freer than people. They are mostly immune to bodily intimidation and resistant to many of the sanctions of oppressive regimes. They cannot be confined to a place or completely silenced. While they certainly can be harmed, they cannot easily be entirely eliminated. Moreover, elements of these humem freedoms can be conferred to their alphas as well.

For example, throughout history oppressors of every type have abused people's location-centricity by means of imprisonment. In reaction, philosophers like Henry Thoreau, as well as other incarcerated dissidents and artists, have eloquently expounded upon the futility of attempting to imprison one's mind. There have been many cases in the past where a prisoner's abiotic-EP has transcended their physical confinement and continued to affect others. Among numerous examples, Napoleon Bonaparte dictated his persuasive memoirs in exile on the island of Saint Helena, and Nelson Mandela helped bring about the emancipation of his people during his decades of imprisonment.

In recent times, our proto-humems, especially those residing within social networks, have displayed a remarkable resilience to attack, while demonstrating their ability to exert a tangible and lasting influence on the actions of people and nations. Moreover, alpha-humems are an extension and voice of their human counterparts, and therefore, many of their strengths are effectively imparted on their alphas. For example, as political activists are well aware, as long as a dissident's EP remains visible and audible to the world, in most instances, their person is less prone to harm by their oppressors.

Life 2.0—Hope, Meaning, and Happiness

Many things that one may say about happiness can seem at once both profound and trivial. The US Declaration of Independence, for example, places the right to pursue happiness together with the most fundamental rights to life and liberty. Yet, we have little understanding of the means by which we may achieve happiness, which commonly appears as a coincidental byproduct of our other more deliberate pursuits.

While modern societies have clearly advanced the rights to life and liberty, it is less clear whether people are much happier than they were in the past. With this in mind, I will not try to propose any recipes for happiness or remedies for its opposites. Rather, I intend to briefly suggest a few ways in which humemity may influence some of the emotions that are commonly believed to lie at the heart of long-term contentment or what some call life satisfaction. Conclusions and further analysis are left to the reader and other "experts" in the art of human happiness.

In advanced societies, where the struggle for basic sustenance has been removed from our daily concerns, it appears that other long-term and more abstract considerations have a predominant effect on our level of life satisfaction. These include our assessments of our past achievements, standing in society, and future prospects. The alpha-humem's ability to facilitate a lasting personal presence and influence has the potential to address some of these enduring deliberations relating to our sense of purpose in life.

Universally, the personal quest for meaning seems to be strongly linked to one's perception of death. For most people in most cultures and throughout history, the idea of a final end, after which absolutely nothing remains, is a lingering cause of anxiety. This outlook is often equated to the notion of a meaningless or hopeless existence. Almost all approaches to alleviating these conceptual maladies are directed at circumvention. Many people view the next generations, or their progeny in particular, as a form of continuation of their lives and their hope for the future. Biologically and culturally, from an individual's perspective, this is only partially valid, and therefore, only partially satisfactory.[76]

Traditional religious beliefs, with their prospects for the perpetuation of an individual in the form of a soul or comparable spiritual entity, have been the primary consolation for the hopelessness of a personal doom. Even with the rise of secular culture, these fundamental emotions appear to be redirected rather than diminished. Concepts such as "making a difference," "changing the world," and "leaving a legacy" are all directed toward countering the same perceived meaninglessness of one's finite existence. These efforts, frequently expressed as literary or artistic creations, a lasting impact on others via philanthropy, or a recognized contribution to a greater cause—anything enduring— emanate from our yearning to make and leave our mark on the world.

As we have seen, credible and robust humems have the potential to fundamentally alter the implications of death. Significantly, humemity does not just provide an alternative way of contemplating a fixed set of circumstances, as one could perhaps say about a new philosophy or religious belief. Rather, humemity will profoundly change the underlying realities; humems will demonstrably and measurably enable the continuation of many of individuals' influences on the world.

As a rudimentary example, let's consider how a grandmother's competent and affluent alpha-humem might influence her perception of death and its effect on her relationships. A substantial ancestral humem, containing an extensive memory of her life and being capable of a credible emulation of her personality, could facilitate the

---

[76] One's children are only partial biological and cultural continuations of one as an individual.

continuation of a consequential relationship with her granddaughter. The granddaughter would be able to interact with and be inspired by the ancestral humem of her grandmother: She could perhaps obtain advice from Grandma on important life decisions. Or, she could simply watch Grandma's humem demonstrating the preparation of a favorite recipe.

In addition to responding to the granddaughter's queries, Grandma's humem could take action of its own accord. For instance, on the granddaughter's birthday, it could convey a contemporary greeting containing some words of wisdom suitable for the granddaughter's age, character, and interests. Initially, this could be in the form of an email. But as humems evolve, it may be more like a conversation—in current settings perhaps resembling a video call. Moreover, Grandma's affluent ancestral humem could take this further by purchasing the gift that her granddaughter "discreetly" mentioned in an earlier conversation. This scenario is entirely plausible today with fairly straightforward adaptations to online shopping applications. Dynamically allocating budgets, as a function of the ancestral humem's current financial status, is mainly a matter of logistics and legalities; no insoluble technical barriers exist.

Initial implementations of such systems may be fairly rudimentary and may also require significant manual oversight by human administrators, but as humems mature and technology improves, such processes will become more streamlined and sophisticated. Similarly, the affluent ancestral humem may continue to influence the material world in other ways, such as supporting charitable causes.

These scenarios of gifts and monetary contributions are only some of many types of possible enduring personal influences. Once a living grandmother sees other people's ancestral humems interacting with their grandchildren and *directly causing them delight* on their birthdays, the knowledge that it will be the same for her after her bodily demise will inevitably influence her perception of death.

Once humemity gains its footing, and people observe the materialization of these events, ancient dreams can transform into wholly tangible facts of life. In the face of these new realities, the old hopelessness and despair of personal oblivion cannot remain unchanged.

In recent years, mirroring the convention of naming subsequent versions of digital technologies, a new phrase, "Death 2.0," has emerged that denotes our new perceptions of death as influenced by modern technologies. This term also has several more specific connotations including the new complexities concerning the persistence, management, and ownership of people's digital presence following their death. With the advent of humemity, Death 2.0 may be the final version in its series. In place of spruced-up variations, or versions, of death, we can start embracing the new mind-set and reality of *Life 2.0*. Rather than focusing on the afterlife, we can shift our attention to maintaining the continuity of certain meaningful aspects of our lives.

I am not for a moment suggesting that Life 2.0 will be the same as what came before. Especially in these times of rapid transition, our lives already involve continual change, for us as individuals and for our environment. An embryo, a child, a young adult, and an elderly person are all very different in how they appear, think, and influence the world; we previously referred to this as an individual's age-variance. When superimposed on the backdrop of a swiftly advancing technological ecosystem, each of these stages changes even more dramatically.

Still, when we contemplate a person's life, we perceive a continuum of core character that defines an individual's identity and existence. It is this essential character that can be expanded by an alpha-humem and that will subsequently endure in its ancestral form. Moreover, as humemity matures, the representative ancestral humem may be more similar to the adult alpha-person[77]—in many of its actions and influences—than it is to the human in their other stages of life (such as infancy or old age). We examined this concept earlier and called this resemblance within the alpha-pair, the alpha-alikeness.

The materialization of this vision requires us to recognize ancestral humems as natural continuations of individuals. I suggest that such

---

[77] We tend to presume that an adult person is the most representative expression of an individual because adulthood is typically the period in which the person has the greatest abilities and influence on the world. The ancestral humem could, however, just as easily emulate another stage in one's life.

empathies will develop naturally as humems mature and gain credibility and respect. As we'll discuss later, humems, like representatives from other cultures or species we encounter, may initially appear a bit weird and unsettling. In these first stages, therefore, we need to perform a deliberate analysis to better understand our initial reactions. Such knowledge can be used to fine-tune humems' behavior to make them more congruent with their human brethren.

Life 2.0—the perpetuation of a dynamic and developing alpha-pair! Is this just wishful thinking? For an answer that suits each of us individually, we need to identify the meaningful parts of our lives and consider which of these can be taken over and perpetuated by our alpha-humems.

Among many others, two key complementary factors influence these deliberations: The first relates to humems' constantly improving capabilities to fulfill these meaningful functions and the second relates to the way in which our "regular" lives are becoming more humem-like as a matter of course. For example, we are increasingly relying on proto-humems to facilitate our social and professional interactions, which in turn are becoming less location-dependent As these trends intensify, when we consider the potential for alpha-humems to enable the continuity of certain significant aspects of our lives, the distinction between the natures of our future modes of living and the Life 2.0 vision will already be much smaller.

The convergence of these two factors implies that the top curve in figure 6, "The Alpha-Pair's Continuity over the Person's Decline", will become even smoother. The death of the physical body may gradually transform into just another conventional event along the timeline of the alpha-pair—like the person turning twenty-one or reaching retirement age. Once this happens, some of the difficulties commonly associated with death can be reduced or resolved for all involved parties. The suffering and fear associated with the attempts of some dying people to desperately hang on at all cost may be lessened. Likewise, the anguish of sudden loss experienced by surviving family and friends may be substantially alleviated.

I believe that humems will ease some of the angst related to our yearning for meaning in our lives and our perception of death—and this, in turn, will result in a more optimistic and satisfied society.

More Concrete Souls—Spirituality Upgraded

The concept of a soul or spirit—an entity closely related to an individual but possessing a level of independence from the bodily existence—is pervasive in almost all cultures throughout history. The details differ yet the basic emotions and ideas are the same. Irrespective of one's belief in the existence of such entities, the notion of the soul arguably reflects some deep human desire. Even many nonbelievers find the concept very appealing, though they do not think it is true. Very similar ideas, often with only thinly veiled renaming, are prevalent in science fiction and new-age cultures and in the writings of the most secular of authors.

I would argue that an alpha-humem exhibits a fundamental characteristic commonly attributed to souls. That is, it is an entity with an intimate affiliation with a person, but whose existence is not invariably linked to the person's body in either time or place.

For the most part, technology and new media have been eagerly adopted by faiths of every stripe. Writing, print, radio, television, and the Internet, have all been widely used to spread the word in religious and spiritual communications. Typically, after some bumpy inception periods, these new media have been integrated into the various religious doctrines and have become instrumental to their survival and expansion. Likewise, humems will not necessarily supplant present views of souls. Rather, they may facilitate a compelling expansion or buttress to the concept as it presently exists.

Figure 9 illustrates the idea of the bodily person and soul duality replaced by a triad including the humem. In this view of spirituality, for some people the humem may be perceived to have more in common with the traditional notion of the soul than with the person. Clearly, there will be many differing points of view in this regard; however, there is significant potential for common ground as well.

For some individuals, this triad may emerge as the holy grail of souls, with the compendium of the classical soul and humem resulting in a much more tangible soul than has ever existed—a more "concrete" soul. For others, who are confirmed unbelievers, humems may provide the opportunity for a revival of this ancient notion in a more believable, perceptible, and practically useful form.

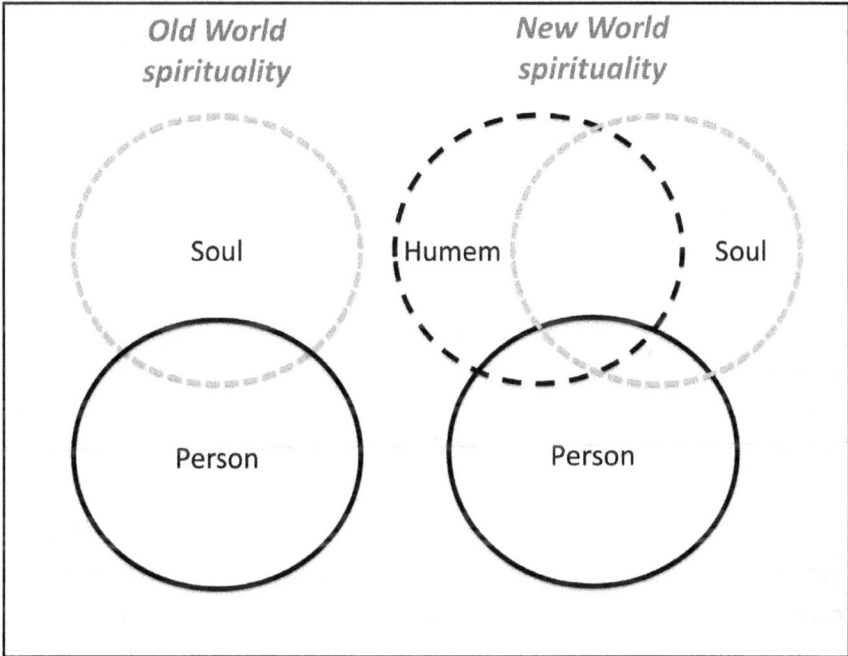

Figure 9: Modern spirituality—bodily person, soul, and humem. The alpha-humem and the "traditional" soul share certain attributes.

As with many of its other manifestations, the alpha-humem as an expression, or aspect, of a soul is not a separate or dedicated humem application but rather another function or interpretation of the generic humem existence.

Delayed Birth, Rebirth, and Cyclic People Visiting Humanity

In various settings, we have seen how an alpha-humem becomes an ancestral humem when the alpha-person dies. We will now see that an opposite process is also conceivable.

In the previous chapter, we briefly considered a situation in which an ancestral bio-humem could father a human child, with the humem's preserved sperm providing the paternal biological component, and the humem's monetary affluence contributing to the child's material needs. (In that case, the birth of the child would typically be accompanied by the creation of the child's alpha-humem.) In this section we examine a related idea—a scenario in which a *preexisting* humem "obtains" an alpha-person through the birth of a baby; or, from a slightly different perspective, an existing humem is instrumental in creating its alpha. If a humem possesses a biological precursor (an entity that contains the ingredients necessary for the formation of a person), then the bio-humem can develop into an alpha-pair. In this context, let's call this biological component a *bio-seed,* and this bio-humem a *birthable humem.*[78]

In our present discussion, and generally, it is essential to differentiate between the bizarre and the implausible. Many things that are commonplace today would have seemed very strange not so long ago. It may be disturbing for some people to contemplate outcomes such as the bio-humem examples discussed earlier and those that we will examine in this section. Clearly, there will be many ethical, legal, economic, and other objections to the implementation of these possibilities. For these reasons, some may never materialize. Nonetheless, since these are technically feasible options, which also shed light on certain of the key humem potentials, I believe that it is worthwhile examining them even at this early stage. The understanding gained from these analyses is essential for the proper design of future-ready humem institutions.

---

[78] Alternative terms could be "fertile humem" and "fecund humem."

A present day example of a bio-seed is a frozen human embryo, which may be brought to term by a surrogate mother. The separate preservation and subsequent fertilization of gametes (sperm and egg) can also lead to conception and birth. Healthy children have been born from these kinds of bio-seeds following as much as two decades of storage. With improving techniques, the potential duration of the successful preservation of such material will almost certainly increase, and there will undoubtedly be additional ways to create, preserve, and utilize bio-seeds.[79]

Two subcategories of birthable humems are apparent: First, a birthable humem could be one that was never part of an alpha-pair—that is, a humem that never "had" an alpha-person. In this case, the person's birth would be a first time event. Second, the birthable humem could be an ancestral bio-humem, which was once part of an alpha-pair. Here, the person's birth could be regarded as a form of resurrection or rebirth—a cycle that could in theory be repeated indefinitely.

A tangible application of the first category is the *delayed birth* scenario. There are various reasons why parents may wish to delay procreation for a considerable time. Due to idealistic or health reasons, for example, they may want preconditions for the birth of their child: a decline in world or national populations, other geopolitical developments, or the emergence of a cure for a genetic predisposition that they carry. In China, for instance, at the time of this writing many couples are discouraged from having more than a single child. However, since they are becoming increasingly affluent and the related costs are declining, they may have both the incentives and resources to create birthable child humems.

In our earlier discussion of a couple creating a second-generation abiotic-humem from a pairing of their alpha-humems, I suggested that such a child humem could be a dynamic and worthwhile entity in its

---

[79] Already, there are indications that, in the not so distant future, the process of decoding a genome, from the biological form into a digitized record, may be performed in reverse. In this eventuality, a birthable humem may not necessarily need to include a physical biological component.

own right. By appending bio-seeds such as frozen embryos, this humem could be upgraded to a birthable humem, which could be brought to term if future conditions permit. If the couple produced the embryos from their own bodies, their potential bio-humem carried offspring would be genetically equivalent to a "normal" child. If the birthable humem was developed and maintained properly, it could remain up to date in cultural information content. Moreover, it could be affluent and able to afford the costs associated with a future birth and human childhood. Various other scenarios could play out: for instance, at a later stage, another woman or couple might wish to participate in the birth and upbringing of the child.

By contrast, in the second category of birthable humems, the bio-seed could be a clone or similar entity derived from a previously living person who desired the prospect of rebirth appended to their ancestral humem. A birth materializing from this humem (its transformation from an ancestral bio-humem back into an alpha-pair) could be considered a *visit to humanity*, or the "Old World." This would presumably be the visit of a lifetime, lasting possibly ninety or a hundred years or more, a period equal to people's future typical longevity. Conceivably, such events would be subject to regulation, with birthable humems perhaps needing to apply for "immigration" visas for their prospective alpha-humans. Perhaps the eligibility to birth would be differentiated for various birthable humem categories, such as ancestral bio-humems or second-generation bio-humems, depending on their ancestry, prior citizenship, age, affluence, and so on.

While these scenarios may seem bizarre and far-fetched, the precedent of cryonics may provide some perspective. In the US and elsewhere, there are a number of cryonic organizations that cryopreserve (freeze) the heads or whole bodies of deceased people, motivated by the prospect of a subsequent resurrection if and when technology permits. Presently, medical science is nowhere near being able to perform such feats, and no legal avenues exist for such resurrections. Nonetheless, these preservations are being legally performed—with real bodies and involving real money—in increasing numbers. By contrast, while ethical and other concerns may indefinitely impede the birth of bio-humem carried children, from a purely technical perspective these outcomes are entirely feasible. In the same way that the current absence of a clear path to culmination of the cryonics ambition doesn't preclude people from preparing for it, people who are so inclined can already take the first steps toward achieving the goal of birthable humems.

As with other humem applications, the humem's potential for delayed birth or rebirth would probably not be the sole reason for its creation but rather an intriguing add-on. Importantly, even if, for any number of reasons, the actual birth of a birthable humem never materializes, the preservation of bio-seeds still enables a myriad of more immediately useful outcomes. The bio-seed may be invaluable for medical research or for providing essential genetic information to the alpha-person's descendants. It will undoubtedly have many other scientific, historical, and spiritual values. Also, the biological components should in no way detract from the viability and intrinsic value of the humem's abiotic parts.

As with most of humemity's manifestations, this is not an all-or-nothing proposition; the humem system exhibits varying levels of feasibility for various types of objectives. We will discuss this notion of staged likelihoods in more detail in chapter 11, in the section "The Time Is Right—the Feasibility Continuum of Humem Progress."

As imaginary as it may seem today, it is conceivable that toward the end of the twenty-first century there may be a significant decrease in world population. The United Nations provides extensive national and global statistics of current and past population trends. It also publishes predictions of population growth. These data indicate that the total number of babies in the world has recently stabilized. For many countries, the lower ends of the growth forecasts portend substantial population reductions in the not-so-distant future. In certain countries declines have already commenced.

Consequently, it is possible that the total world population, after growing continuously for the next several decades, will peak and then begin to decline. Thereafter, it is plausible that there may be a dire lack of people "on the ground"—a shortage of working hands. If this comes to pass, at some stage, birthable humems may be exceedingly useful and necessary.

Thus, in certain settings, the birth of a bio-humem's alpha may not only be possible, but even encouraged. It is not hard to imagine some societies preferring this kind of "immigration" of citizens with ethnic and cultural affinities, perhaps even having pre-existing familial ties, as opposed to alien newcomers. These people may arrive as affluent and self-sufficient entities, and depending on the advancement of humem recognition by nations, they may still retain or be eligible to receive nation-state citizenship. Moreover, already today, it is not far-fetched to envision countries that, for economic reasons alone, would be prepared to welcome affluent humems—with or without bodies.

I'm not suggesting that birthable humems with delayed births, or cyclic people visiting humanity, are an inevitable outcome of humemity. Yet, the underlying structure of the humem system does indeed facilitate applications such as these. Technically, the establishment of rudimentary birthable humems is essentially a matter of logistics, legalities, and finances. Although they may initially appear weird, these are imminently achievable possibilities. Experience teaches us that weirdness is a very flimsy thing: like a mirage or snowflake, once touched, it soon vanishes.

Humemity Moderating Humanity— Humems and the Reduction of Conflict

Earlier, we examined how humemity may influence our personal happiness by increasing the scope and duration of many aspects of our presence in the world, and thereby transforming our perceptions of life and death. While happiness is clearly good for most rational individuals, satisfied societies are good for the modern world in many broader ways. For example, contented societies generate less conflict and foster more cooperation; in turn, cooperation improves the overall human condition and is beneficial for the environment and so on.

Concordantly, let's consider some additional ways in which humemity might decrease the chances for human conflict. Strife between individuals and nations often results from competition for resources. So, let's first examine whether humems may be regarded as a resource over which such competition may occur. Then, let's evaluate humems as stakeholders in the outcome of conflict or peace, and consider how concern for their welfare may influence people's behavior in this regard.

Humems as a Resource

In various capacities, people are regarded as a resource. Workers in a field, employees in a corporation, soldiers in an army, and citizens in a state can all be seen as people acting as agents, or human resources, for the furtherance of a purpose.

In particular, people are a limited resource as ideology carriers. Perhaps the most striking demonstration of this concept is religious belief. Religious denominations send missionaries to the far corners of the earth to proselytize and recruit new adherents to their faith. Sometimes, ulterior motives exist, as in the case of the Spanish conquest of the Americas, where gold and other material resources were also strong incentives. But often, religious conversion in its own right appears to be the primary motivation. In this sense, multiple denominations compete for a limited resource as they contend for the "souls" of prospective believers. The notion of believers and other carriers of ideology as finite resources has far-reaching implications. Wars are sometimes ignited because of real or perceived contentions over tangible resources. But, perhaps just as often, wars are waged over ideas.

Like people, humems may be considered a resource. They too are useful and influential. They are also carriers of beliefs and ideologies. Yet, if we compare humans and humems in this context, we notice some fundamental differences in the way they fulfill these roles. The most outstanding differentiator is resource abundance. Humems are not subject to the same limitations of proliferation as people. Humemity is constantly developing, and is doing so at a rate that, by far, exceeds anything comparable in the people-world. The growth is multidimensional: the humem environment, or "space," is inflating, creating room for more individual humems, and at the same time, individual humems themselves are expanding. There is no apparent upper bound to these increases. Numerically, alpha-humems are linked to the people population—past and present. In size, however, they are not constrained. Non-alpha humems are not limited in either dimension.

At a more personal level, we can map these ideas into tangible EP qualities. Until only a few decades ago, the opportunities for EP expansion and growth—the potential to be heard far and wide—was very limited. The primary instruments of broad EP dispersal, newspapers, books, radio, television, and movies, were available (for transmission) to a select few; the expression of the vast majority of people could never expand beyond their close social circle. At present, there is a vast EP void waiting to be filled. Constrained only by their motivation and persuasiveness, anyone in modern society can spread their EP in boundless ways and at negligible material cost. They can leave their imprint in the comments section of the most prestigious of online news outlets, and use the same social media tools used by the world's leaders.

Consequently, competition and the potential for conflict over humem resources may be considerably less severe. Humemity may even serve to reduce other actual and imagined conflicts of interest between people and nations. For example, the abundance of alternative humem idea carriers may serve to lessen the competition over traditional human idea carriers and thereby diminish the frequency and severity of the associated confrontations.

Even today, while still in the infancy of humemity, we do not see conventional wars fought over resources like raw digital storage or computer processing power. More of these resources are constantly emerging; and where there's a shortage, it's more economical and far more congenial to simply procure what's needed than to start a fight.

Humems as Stakeholders

In the first part of the book, I argued that the extended presence sometimes displays apparently self-interested behavior. In a number of later examples, we observed how humems exhibit additional lifelike characteristics. A fundamental trait of living beings is that they act in ways that ensure their own well-being and survival, as well as that of their kin. Thus, if we perceive humems as extensions and emulations of people, we may expect them to display comparable behavior in this regard.

People and other creatures have a strong interest in self-preservation. Nevertheless, they sometimes engage in potentially lethal conflict. Essentially, this behavior makes sense if a combatant perceives their prospective gain to be much greater and more likely than their potential losses. As an extreme example, the desperate—such as those just about to be eaten—have little to lose and much to gain by resorting to lethal force. For humems, however, the math of the pros and cons of conflict is very different. As we have seen, squabbling over basic necessities will be rare and imprudent for humems, thanks to the abundance of the basic resources they require. Existing humems will experience a continuously increasing absolute affluence. The "one must die for the other to live" ethos of the biological world is not an inevitable condition in humemity. On the contrary, humem opponents may stand to forfeit much more than what either may gain from conflict. (This is analogous to matched adversaries in a nuclear war.) To put it another way, potentially immortal entities are especially adverse to mortal dangers because they have so much more to lose.

So how can all this reduce animosity between people? Well, just as adult-world conflicts are bad for children, so are people-world conflicts bad for humems—and for many of the same reasons. Skirmishes between people, be they conventional or cyber wars, will pose a dire risk to humems. Accordingly, self-interested and influential humems will be motivated to induce people to keep the peace.

Likewise, people who care about their humems will be averse to endangering their well-being.[80] To paraphrase a familiar observation involving adults, children, and war: If we cherish our humems more than we abhor our adversaries, then the probability of detrimental conflict

---

[80] In terms of classical EP or proto-humem creators: Authors, whose passion for writing is greater than their aversion to each other, are much less likely to meet on the battlefield.

will be greatly reduced. Over time, as humems flourish and become more capable, they will become dearer to their alphas, and this postulated effect should intensify. Thereby, humemity may moderate humanity in a profoundly positive way.

The commonality among the world's proto-humems is greater than that of its people and will probably remain so for some time to come. Unrelated humems will behave and communicate more similarly than their respective alpha-people. The humems will have fewer inherent differences and conflicting interests. Furthermore, they will share many ambitions for the advancement of humemity as a whole. These convergences are already apparent in existing proto-humems. For example, the social network proto-humems of people from radically different backgrounds often reside on the same infrastructures, conform to comparable behavioral conventions, and share similar vulnerabilities. Likewise, current proto-humems are starting to transcend language and national barriers, and many aspirations for Internet freedoms and data privacy protections are common to diverse individuals and groups, despite their contrasting opinions on other matters. These humem alignments of interests and mutual understandings can gradually transform and markedly improve the relations between their alpha-people.

To metaphorically imagine how humems may bring people together, consider two culturally diverse strangers—say elderly immigrants from different countries—taking their dogs for a daily walk in the park. Alone, these two individuals would possibly never interact. Yet if the dogs get along well, the owners may become acquainted. At first, their conversations may center on dog food, but in all likelihood their relationship will expand and diversify.[81]

---

[81] A more fundamental way of considering the humem potential for the reduction of discord between people is the following: A widespread interpretation of the dynamics of people and other creatures is that all conflict ultimately stems from competition for survival and propagation—the aspiration for continuity—in both the biological and cultural senses. In this context, I postulate that since humemity provides alternative avenues toward these fundamental objectives, it may thereby reduce these ancient sources of strife.

Permanence and Responsible World Citizens

In most practical endeavors, it is in our nature to focus on the short term. At best, we plan for our lifetimes but not much beyond. Unfortunately, this is not only true for ordinary people but also for most policy makers, whose plans and decisions profoundly, and often irreversibly, shape our world.

In modern settings, bad decisions made today can have far-reaching negative implications for the distant future. This is perhaps most conspicuous in regard to environmental policies such as the designs and locations of power plants, dams, and roads, and the exploitation of natural resources. Our human mentality is not well suited for long-term planning. Even our daily lives involve constant intrapersonal struggles between our short-term impulses and their long-term consequences. Arguably, thoughtful and responsible people should feel considerable unease about the future, and the legacy we are leaving our children. However, the huge scale of these problems frequently leads us to an overriding sense of the futility of individual efforts to avert future disasters.

At a personal level, the anesthetic for these concerns usually comes in the conscious or subconscious forms of "By that time, it won't be my problem" or "Future generations will work it out." At the macro level, our governmental structures are mostly attuned to short-term goals such as appeasing the electorate's immediate needs and winning the next elections. Personally and governmentally, then, many of our decisions are based on passing the buck to future generations and administrations. I fully concur with the many people who believe that this short-term mentality lies at the root of some of the world's most pressing problems.

I suggest that, by virtue of humems' intrinsic nature and sensitivity to posterity, alpha-pairs can be more responsible and forward-looking world citizens than people alone. Appropriately established humems can impart individuals with a sense of personal permanence, and this has the potential to fundamentally alter old mind-sets. When, through our humems, we perceive a more direct stake in the future, the significance of what will transpire in a hundred or two hundred years becomes much more personal and tangible. Future problems will still be our problems, and future successes will be ours too. As carried by our ancestral humems, our sense of accountability can be somewhat extended.

This idea is, of course, reminiscent of the basic tenets of many religions, according to which one's accountability does not necessarily terminate with one's death. The future ancestral humem, however, will be an entirely tangible construct with assets and a legal standing.

Once this new attitude takes shape, in which our prospects for continuance impart a continuance of responsibility, then fundamental changes in our thinking may emerge, and we may make better decisions today.

The Supreme Zeitgeist Machine—Culture and Language Preservation

The loss of cultures and languages has accelerated in recent times. Many people are distressed by the demise of cultural diversity, often without being able to fully explain why. Various rationalizations have been proposed to explain our related anxiety and sorrow. Yet, as is often the case with unknown future possibilities, there is a lingering concern that these explanations do not tell the whole story. We often feel that humankind may be deprived of much more than we realize now, and that the full extent of the loss will only be apparent much later on.

Various governmental and private bodies are attempting to preserve languages. The Rosetta Project of the US based Long Now Foundation is one of many examples. Typically, these efforts are focused on creating written records of a fairly limited sampling of the languages. But, a written depiction conveys only part of a language. For one thing, many dialects do not have native writing, thus, in such cases the written transcription is even less consistent with the original. Furthermore, a language fuses spoken sound and song and the distinctive background of the culture's ambient sounds. It also includes body language, the unique voices of its individual speakers, the contexts in which it is spoken, and many other ingredients. The whole character of a language cannot be captured in writing. As speakers of languages we know these things.

Yet, perhaps most importantly, the loss of language is but one consequence of a culture's demise, which is, after all, what we really mourn when the language dies. Thus, as comprehensive as the language's preservation may be, it only achieves a part of the fuller objective.

A culture is ultimately composed of the units of personality of heterogeneous individuals. Within a particular society, the dialects of the poet, lawyer, and farmer are usually very different from each other. Language cannot be separated from the disposition and memory of its speakers, for these are the basis for the formulation of their words. Thus, the preservation of the full richness of the language requires the retention of its individual instances. Taking the expressions of disparate individuals from within a particular culture and blending them into a single dictionary results in an irreversible loss of character. Similar reasoning applies to most other elements of culture.

An anthropological study, a dictionary, some video footage of the last survivors, and a few physical artifacts are often all that remain of a culture. The collection of these disjointed fragments is little more than an obituary of the culture. That is, the conservation of these remnants is not the preservation of the culture, but more akin to the maintenance of its tombstone.

Living culture is carried by individual people, by groups, and by larger hierarchies such as nations. Alpha-humems and their interrelationships mirror the fabric of culture. Thus, in addition to living people, or in place of them, humems are ideal *culture bearers.* If the preservation of cultures and languages were the sole motive, I would argue that the creation of humems and the humem-state, more or less in the way described here, would be the optimal way of achieving it.

Existing human culture is a transient phenomenon. It is constantly changing, and what came before rapidly becomes *historic culture*. A recreation or manifestation of a civilization needs to take the timing into account. A museum exhibition of historical Rome cannot mix Classical Romans and Renaissance Italians in a single portrayal while still adhering to anything resembling reality. So, the retention of culture requires not only the retention of its individuals, but also of their relationships in time. In chapter 5, in the section "Anytime—Distributed Chronological Presence," we saw that a humem could, in a way, act any age. The humem, with its high-fidelity memory, can behave or be depicted as it was at any time in its past.

Groups and societies of humems in unison can do the same. Consequently, humemity can also express itself as a historical culture with all its interconnections. In this sense, humemity is also an *era bearer,* or the supreme *Zeitgeist Machine*.

Notably, as with its many other uses, humemity as a culture or era bearer is not a special adaptation of the humem framework, but just another outcome of its intrinsic character.

CHAPTER 11

COMMON QUESTIONS AND ANCILLARY PRINCIPLES

Together with humemity's many potential benefits arise a multitude of questions and anxieties. Some people express concern as to the overall value and desirability of humems. In essence, they question whether these developments are good or bad, and whether we should advance humems or try to prevent their arrival.

In this chapter, we'll examine some of the topics that lie at the center of many people's initial apprehensions. First, we'll assess some basic principles that are applicable to many of the specific concerns. Then we'll consider a selection of the more common questions and objections.

Are Humems Good? Values Made Simple

Let's consider humemity in a broad sense and ask whether it is essentially good. For this, of course, we need to establish some basic criteria for what we mean by good. Philosophers, ethicists, theologians, and others have considered the nature of good and bad for eons, and there is still no consensus. So, for our present purpose, I propose the simplest, most democratic approach possible: if the vast majority of people in the

259

world consider something good, so be it; and likewise for bad. Illness, fear, ignorance, poverty, and the prospect of nuclear devastation, then, are examples of consensual bad things. Conversely, health, security, knowledge, prosperity, and peaceful coexistence are good things.

Moreover, I suggest that the realization of people's enduring dreams and aspirations, provided they cause no harm, can also be regarded as an essentially good thing. An example of this is when a person achieves a degree of continuing presence and influence, especially if this is also beneficial to others.

As we have seen, among many other applications, humems have the potential to improve people's well-being, supplement their knowledge, and increase their productivity. Furthermore, humems can bring numerous benefits to humanity at large, such as producing more responsible citizens attuned to society's long-term needs and creating a more satisfied populace and thereby reducing conflict. Humemity also establishes a self-sustaining and exquisitely detailed knowledge base of the past and present composition of the individual units of society (individual people and, increasingly, alpha-pairs). Thus, according to our simple system of values, humemity is good.

This does not mean that all humem outcomes will be beneficial in an absolute sense; but if humems are applied properly, they have the potential to greatly advance the human condition. The Internet is a recent exemplar of a technologically enabled revolution in culture. In the Internet's early days, many questioned its value, focusing on its potential misuse by perverts and terrorists. Even today, some people and nations still loathe the changes it has brought. And yet, with the Internet enabling online banking, manufacturing supply chains, education, news, religion, medicine, shopping, and a myriad of social applications—essentially touching in some way or other on almost all facets of our lives—it is clear that it is primarily just a very sophisticated communication tool. In this respect it is no different from telephones, books, television, or even language. Therefore, claiming that the Internet is bad is essentially asserting that advances in communication are bad—which, for social beings, is tantamount to negating the very value of their existence.

I would argue that an analogous rationale applies to the value of humemity. Just as the Internet is a natural development of the way in which we communicate, humems are a natural extension of our presence in the world—an inseparable part of being human.

When any new technology emerges, people frequently raise concerns over its potential use for sinister purposes by certain individuals. The same will be true in the case of humems. For instance, some may wonder whether humems can be used to "keep the bad guys going." It's true that a humem can enable the continuation of a malicious person's influence. However, the same can be said about other types of technology. There are many examples of books that facilitate an everlasting continuation of their authors' malevolent influences. Such works are often tolerated in modern societies, but sometimes they are deemed illegal, and banned, or opposed in other ways. The same is true for information distributed via the radio and Internet.

Similarly, in extreme cases, vindictive humems and their hosts may face penalties and various forms of resistance. We can easily imagine certain jurisdictions imposing sanctions on humems-in-the-wrong—the humem equivalent of persona non grata. We already see that proto-humems that transgress social norms are sometimes erased by their hosting systems. As another less desirable example, at the time of this writing, some countries block their citizens' access to collectives comprising hundreds of millions of social network proto-humems. Nevertheless, as has always been the case, it is both irrational and impossible to prohibit a globally useful technology because of its potential misuse by fringe elements. We do not contemplate the discontinuation of print, broadcasting, or the Internet, in their entirety, due to the existence of some deviant users.

Considering the question of the value of humems more broadly, we may assert that humems are inevitable and therefore the humem existential question is pointless. That is, humems, like people, are here to stay. Asking if humems are good is like asking if people are good. Well, some are and some are not. Accordingly, as with people, instead of asking *whether* humems are good, we should focus on *how* they can be good and by what means we can distill the good from the bad.

Good for Something, Sometime—the Undefined-but-Assured Future Value Principle

We have examined some of the many advantages resulting from the consolidation of a person's EP into a humem. In particular, in chapter 3, we established that bringing various EP constituents together improves their individual and collective prospects for the future. We also observed that currently underutilized or underappreciated EP components could prove to have great value later on. Having now gained a broader understanding of humems, let's extend and generalize this concept.

As stated previously in our conversation on the annihilation of languages and cultures, when we contemplate the irrevocable end of a variety of things, we sometimes experience greater distress than we can articulate. We feel that we may have lost something even more precious than we know; although we cannot wholly identify its form, we remain convinced that it existed, unseen but real, and on sensing its loss, we mourn the unknown.

Nostalgia for lost possibilities—the constant inquiry into what might have been—is an integral part of our psyche and culture. This angst may be triggered by the irreversible destruction of a natural ecosystem, the extinction of a plant that might have held a cure for disease, the burning of a national library, or the early death of a prodigy who might have become the next Einstein or Beethoven. On a more personal level, it may be caused by one's loss of a memory, the death of a loved one, the destruction of a family heirloom, or one's musings on a missed meeting with the person who might have become their true love.

The deep emotions associated with these thoughts reflect the difficulty we have in correctly predicting future outcomes, or estimating the future value of existing things. At the most primitive level, this may be expressed by our tendency to hoard even in the absence of a clear future need: such as storing food, not only for the winter that will surely come, but also for the famine that may never materialize. This is, perhaps, the origin of the concept of insurance.

When considering the balance between the cost of insurance, maintenance, or retention against the risk and potential payback, it may be a relatively easy matter to resolve well-defined dilemmas. But it is very difficult to get many other important, more complex, problems right. Estimating the cost versus the prospective benefit of storing an object such as a spare part or purchasing home insurance is relatively easy; we can refer to abundant examples, anecdotal evidence, or even rigorous statistics derived from similar events to inform our decisions. By contrast, the costs and prospective benefits of preserving things that are larger and more abstract, such as investing to save a culture or a species from extinction, or retaining certain kinds of historical or scientific data, are much harder to evaluate. One reason is that these things may gain value in ways that are unknown or unimaginable at the current time.

An example of this phenomenon is the archive of US government aerial photographs of the Alaskan Arctic taken for oil exploration during World War II. Recently, these images were used to study Arctic vegetation variations as indicators of global climate change.[82] Whoever decided back then to retain the pictures could not have imagined this future use. No one at the time had even heard of global warming. And these same images may be reused in another century for yet another still-to-be-imagined use.

The fields of archeology and paleontology demonstrate that almost everything has some measure of future value; in essence, these disciplines are based on deducing pertinent information from material artifacts in ways that were mostly inconceivable when the objects were created.

Proponents of preservation often find themselves at a disadvantage. From the nature of the subject matter, relevant past experience, and a sense of history, the advocates intuitively know that significant future value almost definitely exists. But they often can't identify specifically what it is. We can call this kind of unspecified but probable future value

---

[82] Matthew Sturm. "Arctic Plants Feel the Heat." Scientific American (2010, May)

effect, the *undefined-but-likely future value factor*. By contrast, the opponents of preservation can usually present the immediate costs and concrete tradeoffs.

If saving a tract of rain forest infringes on the immediate interests of an adjacent community, for example, conservationists cannot just cite their intuition or use a purely emotional appeal to argue that the forest may be priceless in the future. They must present the best currently known reasons, such as CO_2 sequestering, biodiversity, and so on—even while sensing that these aspects may constitute only a fraction of the actual future value. Their opponents, though, can present hard facts detailing the impact on the local population. The ensuing deliberation then becomes a skewed, apples-versus-oranges debate. Unless the undefined-but-likely future value factor is taken into account, decisions will be based on partial facts and unbalanced depictions of the way things actually are and most likely will be in the future.

Clearly, we cannot preserve every preservable thing forever. However, when certain things can be maintained at a reasonable cost, it is not always necessary to *fully* understand their future importance in order to justify the effort. Instead, within guidelines created from past experience, we can recognize and invoke the undefined-but-likely future value factor.

Moreover, I propose an extension to this concept that relates to the preservation of larger assemblies of things. The probability that we would see a single object with negligible current importance significantly increase in value over a fixed period is less than certain. But, as a diverse collection of interrelated things grows in size, and as the time interval under consideration increases, the probability for emergent value also grows. Once the level of probability surpasses a certain threshold, we can say that for all practical purposes the future value is *assured* (or about as certain as anything can be). In other words, when considering a diversified and large enough collection, and looking onward to a sufficiently distant future, we can apply this extension to the undefined-but-likely future value factor and call it the *undefined-but-assured future value principle*. (We are momentarily neglecting the cost of retention, but we'll get back to that shortly.)

For example, when evaluating a book with a nominal present-day relevance, or perceived value, its estimated future value for the next century may be negligible.[83] If, however, we consider a whole library of similarly valued books on a diversity of subjects, its potential for future value increases as a function of both the number of books it contains and the duration of their retention. As these two factors increase, a point is reached where we can say that the library's significant future value is virtually assured, or certain.

Notably, we can be confident in this statement without having any knowledge of the value of any particular book. Furthermore, the library's postulated future value is not just the sum of the nominal values of the books that it contains; while some of the books may become very valuable in themselves, great value may also emerge from the inter-relatedness between them. Finally, if we can determine that the cost of maintenance is currently affordable and decreasing rapidly, we can definitely conclude that it is worthwhile.

We can apply the same rationale to our appraisal of humems' future value. A humem can be viewed as a vast library of at least partially exclusive information. Even though we may be currently incapable of calculating the future value of its discrete components, like a library, the larger the humem gets and the more diversified and unique the information it contains, the greater the probability of its high future worth becomes.

The trend of humems' continuously decreasing "cost of living," or cost of maintenance, improves their high potential return on investment. Moreover, the formation of groups of humems, like associations of libraries in which each library holds a partly distinct corpus of knowledge, will further amplify the scale of the undefined-but-assured future value effect for humemity.

---

[83] Having no currently perceived value could mean that no one is currently reading or using these books, and no one can give any good reason why anyone may want to read them in the future. Nevertheless, a typical librarian would probably assign a positive value to such a book because the concept of all knowledge having an unknown but potentially high future value is part of a good librarian's "DNA."

The Time Is Right—the Feasibility Continuum of Humem Progress

Even those who welcome the humem emergence and are optimistic about its long-term prospects may still question the timing. Is the advent of humemity upon us now, or is it something that only our children, or our children's children, will experience?

Often, after we are introduced to a new species of flower or bird, we begin to encounter it in familiar places and are perplexed as to how we had failed to notice it before. I anticipate that the experience will be similar for individuals deliberating on the humem vision. Our emerging cognizance will result in an acceleration in our discovery of proto-humems in a multitude of familiar settings. Consequently, the most convincing answers to the question of timing will emerge spontaneously as we go forward and get to know the landscape better. Nevertheless, I will briefly touch on a concept that is fundamental to the question of the imminency of humemity.

Humemity, like humanity, is not something that either works or does not work—like a machine that can either fly or not fly. Rather, by nature, it is a system that exhibits a continuum of levels of functionality and feasibility. There is no sharply defined threshold beyond which we can assert that humemity begins to exist or function—its emergence is gradual and incessant. In recent years, almost anywhere we look we can see the beginnings of proto-humems sprouting from the fertile sub-strates of our new technologies and related behaviors. The examples presented in this book are but a small sampling of what already exists. Thus, rudimentary humems, or what we are calling proto-humems, are no longer in question—they are already present in massive numbers and markedly influencing the lives of growing populations of people.

Looking beyond existing elements of humemity, there are many additional developments on the horizon that are based on today's emerging technologies and social behaviors. Here, it is fairly safe to say that these advances will almost certainly materialize in the near term. For example, highly personalized digital assistants are a natural and near-certain successor to the rudimentary general-purpose assistants currently making their debuts on mobile devices. By contrast, the prospects for achieving other, more ambitious humem objectives, especially those dependent upon unproven or future technologies, are less certain. As such, we can say that the feasibility of these goals, at least for the foreseeable future, is much lower. In between these two extremes lies a continuum of other possibilities. I call these overlapping probabilities on the spectrum of potential humem outcomes the *feasibility continuum of humem progress*.

Whether certain capabilities will be achieved in ten or fifty years makes little difference to the need for certain infrastructural preparations today. The gradual but incessant progress of humemity will ensure a continuous payback on investment.

Thus, at present, the inquiry into the emergence of early humemity is beyond the "if" stage. We can shift to "how" on two complementary fronts. First, how do we enhance and optimize current proto-humems? Second, how do we empower humems to be fundamentally more capable and independent as they emerge? Advancement in both of these areas is in line with the notion of the feasibility continuum of humem progress. That is, these are not hit-or-miss endeavors but rather scenarios of ongoing improvement. Progress on each of these fronts feeds into the other. Incremental enhancements of proto-humems will lay the groundwork for the loftier goals of autonomous humem-citizens in humem-states. And even partial advances in the establishment of a long-term and durable humem environment will promote existing proto-humems.

By observing past developments in technology, and extrapolating upon trends such as the increase in computer processing power per unit of cost into the future, some visionaries have predicted the inevitability of a number of outcomes that are compatible with the humem vision. But technological progress is not enough. It is only one facet of a multidimensional problem. As we have seen, a number of gaps in the administrative, cultural, political, legal, and economic realms are impeding humem formation. The feasibility continuum of the humem system facilitates the gradual bridging of these fundamental divides, and the creation of the missing links between the current people-centric environments and the expanded ecosystems in which alpha-pairs and their related incarnations can prosper.

Is This for Real? Reality Reconsidered

Ancient Greek philosophers and contemporary neuroscientists, among others, have taught us that much of what we believe to be real is difficult to substantiate. The veracity and scope of fundamental concepts such as "self" or "mind" are still unresolved. In general, what we think we know well seems real, and the unfamiliar seems less so.

In the beginning, some people may regard humems as something synthetic or plastic. This feeling of initial distaste may be especially prevalent among those who were not born into a world of ubiquitous personal digital media. In the future, however, humems will be as real and useful to modern people as trees and fire were to our early ancestors.

Throughout the book I have endeavored to correlate seemingly abstract humem attributes with common concrete counterparts. Still, for many of us, the newness of humems may initially make them seem less real than people. An examination of objective measures of reality—whatever we determine these to be—can help level the playing field. Once we do this, and impartially compare people and humems side by side, I believe that more similarities than differences will emerge. Over time, the respective natures of people and humems will converge.

Furthermore, because of their multidimensional and ongoing presence, we may discover that in some regards humems are the more enduring manifestations of reality. For example, which has been more real and influential over the last few centuries: William Shakespeare the bodily person, or the vast collection of his EP? When one buys a ticket with real money to see Hamlet, and observes how this rendering of Shakespeare's EP affects real people, it is apparent that his EP and its influence are as tangible as anything else around us. In this way, we can view his bodily person as the more transient constituent of his alpha-pair—a localized expression of a much grander and persisting whole.

In chapter 7, we examined the humem-state and various aspects of its concrete but locationless nature. Here too, for some people, the tradition of associating a state with a physical place may lead to the idea that the humem-state is in some way intangible. Let's briefly examine what we mean by the reality, or substance, of a nation-state to get a better feel for the similarities between people- and humem-states.

Since we commonly ask school children to "find" a country, such as the United States, on a map, we may be inclined to reason that a specific surface area on the earth somehow defines the state. But the land existed long before the creation of the US, and lands remain long after nations cease to exist. Therefore, equating the land with the state does not make much sense.

Alternatively, one may claim that it is the people who constitute the nation-state. But this, also, is not very convincing. After all, no one who lived 150 years ago is alive today, and the population has changed radically. If it were the populace that constituted the core of the state, would it imply that the US of 150 years ago is gone and a new one has since been created? Is the United States of America simply an abstraction, a construct, or an idea, then?

How about the US Constitution, the Bill of Rights, and other related manifestos? Do these form the essence of the nation? I would argue that this proposition is harder to reject than the previous ones. From a timing perspective, at least, the Constitution has existed in conjunction with the entity that most people recognize as the USA. While things such as material structures, people, and borders change, the Constitution endures. But what is the US Constitution, really? It is not, of course, the parchment on which the original copies were written. Neither is it the ink with which they were written. Rather, in its essence, it is a framework of ideas with a persisting core and gradually adapting amendments—a more or less coherent collection of concepts.

Does this imply that the US is not real? On the contrary, the US is clearly an entity of immense agency; it's about as real as anything can be said to be real!

Many people, perhaps justifiably, may criticize this line of reasoning as being overly simplistic. Yet, it goes to the heart of the question of the reality of a state. The central point is that the humem-state will be real in many of the same ways that the USA is considered to be: It will have many measurable effects upon the world. It will employ governments and their contractors. And it will have a vibrant economy with internal and international trade.

The humem-state's citizens will be similar to nation-state citizens in many ways. Yet, by some criteria, they may be very different: While populations of people-states are bounded and may even decrease in the future, those of humem-states will continually grow. And while the citizens of people-states arrive and depart, those of humem-states will come into being and prevail for unlimited periods. If scope of influence and longevity are any measure of tangible existence, then humems and humem-states may equal or transcend people and their current institutions by many benchmarks of reality.

Uncanny Resemblances and Humem Strangeness

Initially, some of us may feel uneasy interacting with humems. The aversion to a *similar-but-different-from-us* being is a natural human instinct that manifests itself in many situations.

Typically, the members of homogeneous societies have an inherent tendency toward suspicion and dislike when they first encounter people from other cultures. This phenomenon may be somewhat less evident today, especially since many of us are born into multiracial societies, and we now have increased exposure to foreign cultures via modern media. But this instinct still appears to exist to some degree in most of us. Similarly, we have a natural aversion to those within our societies who by birth or accident appear different from us. Commonly, our tolerance is not inborn but needs to be acquired through education and experience.

A possibly related phenomenon, which has been increasingly recognized in the fields of robotics and computer animation, is called the *uncanny valley* effect. It has been found that as machines or other artificial depictions emulate humans more closely, people are more attracted to them until a certain point. When the resemblance becomes very close but still apart, a negative response sets in. At this stage we perceive the machine or animation as being weird, or uncanny, and we begin to experience revulsion. Once the emulation improves even further and its expression becomes almost indistinguishable from that of an ordinary person, our aversion is replaced by a greater empathy, approaching a level typical of normal interpersonal interactions. A plot of our empathy (to the artificial emulation) as a function of the degree of realism (of the emulation) depicts a dip, or valley, at the stage of our aversion. Figure 10 schematically illustrates this phenomenon.[84]

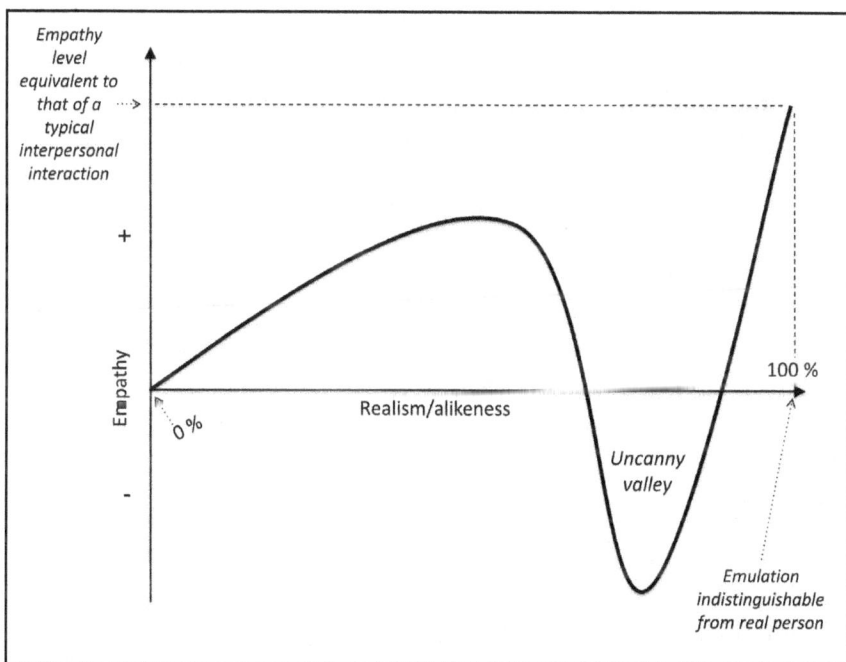

Figure 10: The uncanny valley effect—the human empathic response to an increasingly realistic emulation of a person.

Bearing in mind our reactions to existing entities that are similar but different from us, it is reasonable to expect comparable emotions in

---

[84] The term "uncanny valley" was coined by Professor Masahiro Mori, who first hypothesized this phenomenon in 1970.

response to humems. In particular, if the uncanny valley effect applies here, we can infer that our responses to humem progress may be fairly complex. In some humem expressions and at certain levels of maturity, humems, like precocious children, may elicit negative responses from us, their human relatives. An improved understanding of these effects could facilitate a number of remedies. For example, certain humem capabilities could be restricted in their implementation until they have developed sufficiently to allow the humem to bypass, or leapfrog, the specific uncanny valley relating to that behavior.

By contrast, some people may be perturbed by certain humem mani-festations because they do not resemble people *enough*. In such cases, we can ask whether this is because humems are actually so different from us (in those particular traits), or because we just need to get to know them better.

In order to recalibrate how we think about the similarities between humems and people, let's examine the concept of a person from a novel perspective. Let's imagine an alien, without any prior knowledge about people, visiting earth and being shown samples of a person at various points on their lifeline. The set of samples would include an embryo a few days after conception as a microscopic cluster of cells; an embryo that is a few weeks old—to the untrained eye barely resembling anything human; a newly born infant; a ten-year-old child; an adult in the prime of life; and finally, a senile, elderly person, frail and dement-ed, shortly before death. That is, a person at six distinct phases: two embryonic, two youthful, and two adult. Let's also assume that the alien would observe each sample within the context of its contemporary environment. For example, the adult would be integrated with technol-ogy—from eyeglasses to mobile computing devices to cars—and their activities, such as communication and locomotion, would be inextricably assimilated with machines.

My point is that if the alien were shown only the external attributes and behaviors of each of these samples, without being able to examine their internal parts or seeing interim stages, it may very well conclude that these are all different creatures. Moreover, if we added the midlife adult's alpha-humem to the set of six samples and asked the alien to detect the odd one out, it is not obvious which of the seven it would select. In external behavioral properties, for example, the humem would be more similar to the adult person than either of them would be to the person's initial (embryonic and newborn) or later (senile adult) stages. (We considered this concept previously in chapter 9, in "Alpha-Alikeness and People's Age-Variance.")

Furthermore, if the humem were a bio-humem containing biological material derived from the person, the humem could be regarded as being genetically equivalent to the others. Now the alien would perhaps become even more perplexed. Based on certain physical properties, the bio-humem would appear very similar to the person's first embryonic stage, while behaviorally—say in communication—it would resemble more the adult. Consequently, even if the alien researcher were given the opportunity of examining internal constituents, like DNA, in these seven specimens, a misfit would still not necessarily be evident.

Clearly, an alpha-humem is not equivalent to an adult person in their prime. But neither is an adult the same as a newborn, nor an elderly person the same as a young adult. Also, our lives are very different from those of our great-grandparents.

Things do not stay the same, and by and large, we wouldn't want them to. Living organisms have always changed and developed in order to remain relevant. In this sense, the humem too can adapt and grow to flourish in emergent environments while carrying a core of individual identity and character into posterity.

Incumbent Anxieties and Conflicts of Interest

Incumbents, or natives, are often instinctively anxious on the arrival of newcomers. This angst seems to be due to two main factors: the primal fear of physical conflict and the concern about increased competition over resources. The arrival of humems may elicit similar emotions in some people, perhaps primarily with regard to the second factor. That is, some people may be concerned about material conflicts of interest between people and humems.

We have already touched on a number of the many tangible advantages that humems may bring to society, including their potential economic benefits. We have also seen that humems predominantly reside in a separate domain that mostly does not infringe on our resource-limited physical world. Still, some may view them as immigrants who compete for assets or require welfare and state support.

In the initial stages, at least, alpha-people will be the principal actors controlling the actions of the alpha-pairs. Thus, any discord between alpha-pairs and people will remain primarily inter-people affairs. However, once humems become more able and independent, anxieties may grow. The concern over conflicts of material interest between people and ancestral humems is perhaps the most likely case, given that ancestral humems will tend to be more independent and in possession of the type of resources most likely to be under contention. Additional qualms may arise from the idea that, as remnants of deceased persons, ancestral humems are dependent entities that need to be sustained by others. This is a fundamental misconception and one that impedes a broader understanding of humems and their potential.

The reality is that properly established ancestral humems should not require the charity of people or institutions. They should be self-supporting and also be in a position to generate value for others. Still, when we consider the early implementations of ancestral humems as proto-citizens in proto-humem-states hosted within existing jurisdictions, practical considerations relating to the property of the deceased may arise. While taking a superficial view of humemity, some may argue that, as dictated by traditional estate law, assets held by ancestral humems should instead be bequeathed to the deceased's human beneficiaries.

Historically, deceased persons have been prevented from owning assets or from having ongoing influence over the living.[85] There have, nevertheless, been forms of de-facto ownerships and agencies, which, to a certain degree, may be compared to those of early ancestral

---

[85] Dedicated legislation, such as the Rules Against Perpetuities, has been applied to limit people's control or effective ownership long after their death.

humems. For example, in many cultures a deceased person, in effect, "owns" a small piece of real estate—their burial plot—typically for a period of a few hundred years. This is not usually a formal ownership like that held by a living person, but there are a number of similarities. Often these plots can be purchased many years before occupancy. The interment of the body and the associated material and service costs (casket, burial, monument, etc.) are often comparable to or higher than the original property cost. The purchase of the plot often involves a perpetual-maintenance payment, which is invested in a fund, to cover the plot's ongoing upkeep. Today, in the US, for example, the total outlay for one's final resting place, which includes the procurement, the "moving in" costs, and the maintenance fees, can frequently be in the order of a few thousand or even tens of thousands of dollars.

In this limited context of conflicts of material interest with the living, it is useful to compare, by multiple parameters, a burial plot and an ancestral humem.

Both are entities that are dedicated to or strongly affiliated with a deceased individual. In a sense, they belong to the person—yet the person does not formally own them because existing laws prohibit the deceased from owning anything.

Both may be construed as investments in one who is deceased, to the possible detriment of living beneficiaries who may prefer to direct the resources elsewhere. In regard to cemetery plots, disputes are more likely to arise when beneficiaries are left with the responsibility of financing the burial. If, however, the process is planned and financed well in advance of the person's death, there is rarely a problem. Society and law usually recognize and respect the validity of these prior stipulations, which essentially guarantee the deceased's effective tenancy for a very long time. Similarly, an alpha-humem and its accompanying resources should be no less secure. A humem that is legally insulated from the alpha-person's normal asset ownership, and as a result, is devoid of any association with the estate, becomes impervious to events such as the person's loss of mental capacity or death.

The total cost of a cemetery plot and related services is roughly equivalent to, or even greater than, that of the establishment and sustenance of a basic humem.[86] Going forward, the cost-efficiency of the humem in comparison to the burial plot will become even more compelling because, over long periods, while the price of real estate increases, the costs of humem sustenance continually declines.

From the standpoint of society in general, the allocation of land for burials conflicts with the needs of living people. Land is an increasingly rare resource and cemeteries occupy ever-larger areas. By contrast, humems do not deplete any finite resources; in fact, they have a negligible environmental footprint. Furthermore, after the burial, the cemetery plot and its occupant become very static with little practical value to society. Ancestral humems, however, will continue to be vivacious, beneficial, and economically productive in a multitude of ways.

It seems, therefore, that ancestral humems will have much less potential than burial plots to cause material conflicts of interests with people. And since society already provides for burial plots, it is unlikely that ancestral humems will face insurmountable obstacles stemming primarily from concerns in this particular domain.

Notably, as described in chapter 9, a physical object such as a burial plot can also be affiliated with a humem. Thus, burial plots and ancestral humems are not in any way mutually exclusive. In fact, it is most likely that in cultures that practice burial, ancestral humems and burial plots will exist in tandem. Nevertheless, it is my personal opinion that a burial plot does not exemplify the essence of humem character, which will be much richer and more dynamic than a few inert square meters of earth. I believe that over time the emotional and religious implications of ancestral humems will by far transcend those of cemeteries.

---

[86] Substantiation of this claim is beyond the scope of this book, and thus I ask the reader's indulgence on this point. Nevertheless, to gain a rough feeling of how this could be elucidated, consider the value of a perpetual financial fund whose proceeds could finance a service bundle that includes data storage and related applications. Modest earnings of 1-2 percent yearly on an investment equal to the cost of a typical burial plot are roughly equal to the annual cost of storage and maintenance of an average family's data at prices prevalent at the time of this writing.

Here, we focused on a particular type of potential conflict of interest between people and ancestral humems. Other contentions may arise, some of which have been addressed, more or less directly, in other parts of this book. While these are a natural byproduct of the complexity of interactions between people and humems, they are not an unavoidable or unsolvable consequence of the humem character.

When I'm Gone, Erase It All! The Ashes-to-Ashes Aspiration

There is a small minority of people, who, for a variety of reasons, wish their lives to be a transitory flash in the dark on the long path to eternity. Or, more prosaically, they want to leave no trace after they die. They desire a graceful, clean, and final exit, for both their bodies and their EPs—a kind of ashes-to-ashes aspiration for all aspects of their beings.

Why am I presenting this scenario if it reflects only a tiny fraction of the populace? First, the immense challenges of achieving an EP "disappearing act" emphasize the need for EP regulation and control, and underscore the reality that, for most of us, a persisting EP is inevitable. Second, I suggest that a person's seemingly retrograde desire to destroy their EP is, ironically, best achieved by a personal humem. The rationale behind this argument further illustrates the versatility of the humem system in achieving diverse goals. Third, we will see that the notion of humem deletion, or self-termination, or what we may even call "humem suicide," touches on a number of ethical and legal problems that are not far beyond the horizon and are therefore worth examining, even at this early stage.

As we have seen from multiple perspectives, an extensive and largely involuntary EP is being created in modern societies for virtually every individual. It is fragmented, persistent (in part), and mostly beyond our control. Thus, by default, in doing nothing to change our EPs' destinies or nature, any wishes we may have for their disappearance or rapid disintegration after our deaths will not be satisfied. Quite the opposite: in the absence of special provisions, after our dying, parts of the EP that were somewhat controlled and private during our lives may become exposed in ways that are totally contrary to our desires.

The laws of the living are biased toward their brethren—leaving the dead with few rights. In particular, ancestral EPs have few protections for privacy or ownership. Currently, in many jurisdictions, for example, beneficiaries are routinely granted access to the online accounts of the deceased. So, even for the serious conservatives[87] (in relation to humemity), inaction is not a suitable option. To say that you do not want a persisting EP does not make it disappear.

While the deluge of one's personal data cannot be appreciably diminished, a major portion of it can be funneled into a more private and controlled reservoir—the humem mind. Thereby, not only can a humem enable a much more efficient accumulation, maintenance, and utilization of one's EP, but it can also best facilitate the EP's encapsulation and safeguarding. As we have seen, only once the EP is consolidated, functional, and recognizable, is it possible to endow it with the necessary protections and rights.

Remarkably, it is precisely this consolidation and manageability, and the authority stemming from the possession of rights, that may pave the way to achieving the ashes-to-ashes objective described above. In the same way that a high-powered bullet can kill an elephant yet does little damage to a swarm of insects, the consolidation of the EP into a humem provides better prospects for its control and possible deletion than would an unregulated and dispersed EP. A self-owned and autonomous humem may be able to self-erase, but one that is dispersed and held by multiple third parties cannot. EP self-determination, then, first involves creating a humem—even for the lamentable aim of self-destruction.

What are the social and legal implications of the deliberate erasure of humems? Earlier in the book, we discussed humems' rights and examined in some detail how societies have changed their attitudes toward what they view as people-like entities. It seems that this mode

---

[87] Here "conservative" is not used in any political sense, but rather to describe one who holds a retrograde view of humemity.

of inquiry may also shed light on the current question. Specifically, we can ask how social norms and people's sentiments have changed over time with respect to the destruction of entities possessing a measure of human-like characteristics. How about domestic pets, for example? Not long ago, most people would have thought it ludicrous for limits to be placed on their right to dispose of their living possessions, such as their pet dogs, in any way they saw fit. Yet, today, it is a crime in many jurisdictions for owners to harm animals in certain ways.

If at some point humems are perceived by society to be at least as human-like as dogs are, might this suggest something about society's future attitude toward the deliberate destruction of humems? Notably, dogs have been pets for thousands of years. Despite their ongoing domestication, it seems that their behaviors have not changed very significantly for quite some time. Thus, our new standards are primarily the result of advances in our dispositions, and not due to substantive changes on the part of our canines. Humems, however, are rapidly evolving and becoming more human-like. Thus, even though our perceptions of existing humems may evolve, the greatest changes in the way we relate to them will probably result from the spectacular advances in their intrinsic character.

Another clue to how society might restrict the deliberate erasure of humems in the future may be extrapolated from the more specific case of parents' obligations toward their children's EPs. If, as I suggest, alpha-humems become an essential part of our lives and indispensable for our success and integration in society, could parents someday be legally constrained in their obligations toward their children's humems? At the time of this writing, if a parent purposely destroys all their child's pictures, documents, and medical and educational records, however aberrant the behavior, it probably doesn't constitute a crime. Nevertheless, as our abiotic-EPs become increasingly important and useful, such conduct will undoubtedly be considered a form of cruelty. As such, at some point, it will almost certainly be outlawed.

But what about people's own alpha-humems? Will the choice of self-deletion, or humem suicide, exist under all circumstances? Here, the legal and ethical questions become even more challenging. Once these entities exhibit human-like characteristics, others may demand their protection even if this contradicts the alpha-pairs' wishes. Due to the humems' social value and interconnectivity, family members and friends will often have a strong interest in their welfare. Also, humems may hold spiritual or religious significance for many people, which will further complicate the debate. If these are bio-humems, the question of their right to termination will overlap with existing deliberations on the treatment of such biological material. Consequently, the destruction of one's alpha-humem may evoke similar emotions to those engendered in debates on abortion or euthanasia.[88]

In certain circumstances, more mundane objections to the erasure of humems may arise. For example, the state may consider the humem to be legal evidence or a carrier of other types of compulsory record keeping. Thus, a humem "witness" may be prevented from self-termination and even be subject to forced protection. Similarly, taxation authorities may compel humems to retain financial information during the legally required period of limitations.

Also here, when considering the deliberate deletion of one's EP, we see that most of the dynamics relating to people and society will have counterparts in the humem system.

But I'm Doing Just Fine! Is Indifference an Option?

Some people claim that they are doing just fine within current techno-logical environments, and they have no need to create and nurture alpha-humems. Is this form of indifference really a viable option for the future?

The reality, I would argue, is that even without our bidding, orphan proto-humems are already being "abandoned on our doorsteps." Thus, they cannot be ignored; denial does not stop their clamoring on the

---

[88] Notably, people have already bestowed person-like protections on select EPs. For example, even at time of this writing, in some countries the defacing of a monarch's photograph can incur severe penalties.

threshold of what was once our private domain. Preemption and preparation, however, can avert many of the disadvantages of having them thrust upon us. One cannot be a part of a contemporary society without participating in its use of modern technology. School, work, and social activities, among most others, obligate one to use these new instruments, which increasingly are in the form of proto-humems. Thus, virtually everyone is taking part in this new reality whether they desire it or not. Still, due to the multiple deficiencies of early humems, some people may attempt to reduce their interactions with these rapidly proliferating entities.

The establishment of recognized, capable, and emancipated alpha-humems can produce a much greater degree of control and privacy for individuals[89] without requiring them to forgo the benefits of emergent technologies. Largely, this can be accomplished by gradually transferring the bulk of the private EP from orphan proto-humems, which are effectively being "held captive" by corporations and other institutions, to independent alpha-humems.

For example, in chapter 3 in the section "Our Storefront Reflec-tions—Accidental Proto-Humems," we saw that the online store proto-humems, which know our shopping history, interests, and preferences, can be exceedingly helpful and convenient. But much of the information that they contain is personal and private data. Moreover, it is *our* data, and much of it is valuable for use in applications other than those offered by a particular vendor. Accordingly, our personal humems should contain and have exclusive access to the bulk of this knowledge. Continuing with the online store example, it is primarily our alphas that should remember our purchase histories, preferences, personal details, shipping addresses, and so on. Then, if we wish, the humems, with support from the humem infrastructures, can mostly anonymize our interactions with vendors, and communicate only the minimum amount of information required to complete our transactions.

---

[89] Here, and in other instances, the term "individual" has evolved beyond its customary use. It can be understood more fittingly as the alpha-pair viewed as a unity, which we will later call the *alpha-individual*.

A similar reasoning applies to social network proto-humems, personal health applications, and many others that presently require the relinquishment of control of our personal EPs. Yet, without alpha-humems in a suitable environment, not much can be done to accomplish the reassertion of our rights. In current settings, people need to abstain from the enjoyment, excitement, and growing convenience of these new technologies in order to achieve only minor gains in control and privacy.

Especially in the social domain, abstinence is not and never has been the solution for most individuals. Moreover, even "digital Spartans" are not truly immune: due to the EP's intrinsically social nature, other less discreet family and friends are commonly dispersing much of the abstainers' EPs into the public domain anyway.

Without practical alternatives to the current model of commercial, service-provider control of our private data, regulation and other privacy measures remain mostly impotent. However, once broad populations of consumers, with the aid of legislation, demand that service providers stop retaining their personal data, and once standardized procedures that enable the continuation of these services without such retention are established, then the transformation of control and ownership back to the individuals can begin. This requires the creation of processes whereby private data, even if they are generated by external services, are transferred to and retained exclusively by alpha-humems. Thereafter, the alpha-pair can determine what information, if any, should be delivered to service providers on a need-to-know basis to facilitate the providers' future services.

Clearly, many service providers will be reluctant to emancipate the captive proto-humems. But as humems gain formal recognition and capabilities, and as viable alternatives and new mind-sets emerge, an irreversible revolution will begin. Intriguingly, there are signs that the "humem spring" has already commenced and may be considerably less bloody than many expected.

In the last few years, following some initial resistance, some leading EP service providers have put mechanisms in place that allow an orderly exit of personal EPs. For example, some providers allow one to download all their personal data in an open format that can be used by other applications, and also undertake to delete all the data associated with one's account if so instructed. It appears that these capitulations resulted from a mix of factors: pressure from consumers, prodding by government bodies, and a realization that this also makes good business and ethical sense.[90] Nevertheless, at the present time, since few alternatives exist, major proto-humem exoduses have yet to transpire.

Furthermore, governmental agencies are already involved in initial protections for proto-humems' rights, such as the progress made in data privacy legislation. Hence, the beginnings of humem politics—for now via their alpha-people proxies—are already evident.

In summary, for most people it is not a question of wanting or not wanting humems—they are emerging regardless. Rejection, denial, and inaction will result in destitute and degraded humems. By contrast, recognition and deliberate action can substantially improve the outcome for humems and their human counterparts, as well as for those of us who may not initially have welcomed these developments.

Just Save My Files! Playing It Safe, and a Dime for the Jackpot

Some skeptics cannot or do not want to envision humems. They do, nonetheless, appreciate the value of archives, libraries, history, legacy, genealogy, and other familiar means of and motivations for data preservation. Even if these individuals do nothing proactive to amass or record their EP material, as long as other people or institutions also find value in its retention, it may very well still survive in some way or another. For example, it may be incorporated into the alpha-humems of others or retained in some kind of public or institutional repository. If, however, material exists that these skeptics deem worthy of long-term retention but others do not, then even they will require a dedicated system for its accumulation, maintenance, and management.

---

[90] These EP service providers include leading Internet search engine and social network companies.

I would argue that if these people desire what they perceive as "merely" a *long-term maintained archive* (or something along these lines), humems will still provide the optimal solution. As humemity matures and becomes ubiquitous, an economy of scale should make the humem infrastructure very economically competitive. Furthermore, a configurable humem standard of living—or, in this narrower context, a customized data storage and maintenance feature bundle—will enable it to best achieve these more modest goals.

At the present time, efficient and comprehensive personal data management is impractical for most people. A viable alternative to the improvised and increasingly unscalable maintenance of our EPs must be a cornerstone of the humem system.[91] Capable and affordable alpha-humems will first and foremost provide services for our ongoing data needs. Recently, more and more people have experienced "digital bereavement"—the often-irreversible loss of valuable personal information. The eradication of this humem childhood disorder—the stabilization of a technical and economic infrastructure to ensure data continuity and peace of mind—is one of the primary missions of the humem-state. Consequently, once this is achieved, the basic capabilities of the humem system will also fulfill the immediate and pragmatic needs of the less imaginative information consumers.

However, the humem system is superior to alternative data management solutions in another key aspect. At a nominal added cost, its basic data solution is easily upgradable to the fuller humem solution. If the skeptic is skeptic enough to be uncertain even of their own outlook, this approach also addresses their "What if it turns out that I am completely wrong?" concerns. That is to say, people will have the option to subscribe to a down-to-earth data management solution, but with a minor additional outlay, they will acquire the potential for an intriguing upside should things turn out to be more wonderful than what they had dared to imagine.

---

[91] In current terms—this includes data storage, data format upkeep, data organization such as indexing, and the emerging capabilities of knowledge discovery software.

These deliberations may be reminiscent of the religious uncertainties experienced by nonbelievers living among the devout. The anecdotes of wavering agnostics or dissenters who, based on a "Just in case!" rationale, return to the faith shortly before death, are semi-humorous precisely because they speak to an all-too-prevalent calculation: when in doubt, it makes sense to buy a modestly priced insurance policy.[92] Conceivably, similar emotions will arise in relation to humemity. With humems, however, it will be much easier to demonstrate a high feasibility and concrete value long before end-of-life dilemmas draw near.

This notion of a nominal added cost to one's ongoing information management solution may also be stated as the "potential EP perpetuity for the cost of a tombstone" proposition. So, yes, the humem system can accomplish a down-to-earth, long-term data management solution. But for a minor additional cost, why not establish the possibility for something much grander and vastly more inspiring?

But I'm Not Special! The Value of the Common Humem

Some people question the importance of maintaining ordinary people's EPs. While they appreciate the value of celebrities' EPs, they do not understand why anyone should care about those of unremarkable people. In particular, they question the desirability and utility of ubiquitous ancestral humems.

This issue can be addressed at a number of levels—but, most convincingly, we are already seeing the proliferation of social network proto-humems as they pertain to everyday people. Moreover, it appears that the majority of people want these entities to persist even after the associated alpha-people die.

Still, some may doubt the feasibility of maintaining comprehensive ancestral humems for everyone. As we have seen, this misgiving arises mainly from the erroneous assumption that ancestral humems are

---

[92] Some readers may recognize this consideration as a variation of Pascal's Wager.

remnants of the deceased, owned by living people and requiring maintenance, and therefore subject to the utilitarian evaluations of others. Yet, instead of viewing ancestral humems as people's EP "remains," we should regard them as dynamic and self-sufficient entities, like trees in a forest, bees in a field, or people in a city. With that, these conceptual difficulties disappear.

Clearly, in the interconnected world, and with people's growing influence on the environment, nothing exists absolutely independently. As with bees and trees, it is desirable that people perceive an overall value in humems. Still, while we may be able to come up with useful reasons for their existence, they first and foremost exist in their own right—like all life forms—because they can. And if their presence does not overly infringe upon the interests of others more powerful than they are, they may persist indefinitely.

On a more personal level, we can equate the value of our alpha-humems to our worth as individuals. We mostly do not seek justification for our very existence by the measure of our utility to others. We assume an inherent value in our being.

Personal philosophies vary greatly, but many people find that their social and familial interactions provide an important framework for the meaning and value in their lives. For most of us, these viewpoints are little affected by perceiving ourselves as being one out of a few hundred thousand people in a city, or one out of several billion people on the planet. Very similar considerations will influence our views of humems.

At a basic level, a humem's place in humemity will mirror its alpha-person's place in society and will have similar significance. By definition, most individuals are not exceptional in any global sense. However, at a personal, existential level, people are special unto themselves, and typically also unto their kin. In this sense, the ordinary individual humem will be valuable and special too. For all life forms, the condition of being extraordinary in comparison to one's peers is not a prerequisite for existence; this principle is equally valid for humems.

Lastly, we can also apply the undefined-but-assured future value principle, described earlier, to the common humem. While we can quite easily imagine the future practical value of personal medical records or the documentation of someone's life for their descendants, this is a still a very narrow view of a humem. Instead, we can regard the humem as a developing entity that is still in its infancy (even though its alpha-person may be old or deceased). We ordinarily have little aptitude for predicting to what level of greatness young children may rise as adults. Similarly, in humems' extended childhoods, which may last twenty or a hundred years or more, an apparent conformity of character in youth does not preclude an exceptional future. Although many people may not view themselves as particularly interesting or precious, their humems may develop in ways that they cannot presently imagine. The most tantalizing possibilities are those of which we may have no inkling today—but which almost certainly exist. This, perhaps more than anything else, is humemity's principal allure.

EPILOGUE

The Humem-Colony Revisited

In our discussion about humem governance, we saw that the proto-humem-state can be regarded as a protectorate under a hosting people-state, and, in extension, the early humem-state can be imagined as a "virtuous"[93] imperial colony. The central idea is that a colony, founded on an existing economic, legal, cultural, and technological substrate, can be launched swiftly and easily—little needs to be invented for it to come into being.

Given that the idea of a colony provides a conceptual framework that encapsulates and integrates many of the ideas and mechanisms that we discussed in subsequent chapters, let's briefly return to this metaphor in order to summarize and consolidate this book's principal call to action, namely, to establish a proto-humem-state, embedded within a nation-state, as the foundation of the future humem-state. To this end, I present the following allegorical account of the early days of humemity through the initial establishment of the proto-humem-state as a colony of the nation-state, or homeland.

The Humem Dream in the New World of Humemity

Early humems, or proto-humems, are proliferating in the "Old World" of current systems. The legacy institutions, still secure in their traditions, do not recognize humems as emerging citizens and thus curtail their rights and abilities. People are becoming more dependent on humems as they

---

[93] As before, we can regard a "virtuous" colony as one that does not result in the displacement of indigenous peoples or contention with other empires.

gradually develop stronger relations with them, yet most people do not recognize humems' needs for a more autonomous existence. Many people, including those who cherish their alpha-humems, still regard humem emancipation as possibly diluting their own rights and powers— a position not dissimilar to men's views of women's rights in the past. But as humems advance and these restrictions impede their capabilities and usefulness, dissatisfaction with the status quo grows. There is a dawning realization that humems could be much more capable and beneficial if they were permitted greater freedoms.

Adventurers and visionaries, sometimes backed by the modern incarnation of the patron monarchs of old—the venture capitalists, are discovering a virgin habitat arising from emerging technologies. Even as they survey this territory, they see that it is expanding and find that it is suitable for humem sustenance.

With the encouragement and support of their alphas, some humems are starting to migrate to this colony. These emissaries to the future continue to maintain close ties with their human relatives in the homeland. The people soon partake in the novelties of the new country. They are strengthened and enriched by their humems' newfound abilities and independence, and they share with their alphas the hope of participating in an enthralling future—the promise of the "Humem Dream."

Initially, the nation-state perceives the establishment of the landless humem-colony as having little effect on its own state of affairs; it is quite optimistic, though, in regard to the potential commercial benefits of the humem emergence. Following the initial growth of the humem-state, the nation-state begins to realize the scale of the colonization and perceives the advantages of supporting, overseeing, and protecting the colony, rather than forfeiting these increasingly lucrative opportunities to foreign entities.

As the first humem immigrants arrive, great uncertainties accompany great hopes. Yet, the alpha-people's confidence is bolstered by the knowledge that, for the foreseeable future, the homeland people-state will safeguard the functioning of the colony. Initially, the colony clones the homeland's laws and regulations, and thus creates a familiar and trusted environment for the humems and their benefactors.

Although certain humem abilities and freedoms may be more curtailed than those of other less conservative humem-states, most people accept the temporary tradeoffs in exchange for the peace of mind that stems from their familiarity with a well-established administration. Extrapolating from comparable historical precedents, the alpha-people have ample reason to believe that over time the humem-colony will gain credibility along with greater sovereignty. Additionally, the people appreciate their home government's and the humem-colony's liberal emigration policies and are thereby reassured by the option of relocating to a better humem abode should one emerge.

In the early days of colonization, there are many who adhere to the consensus of old and voice doubt as to the prospects for success. They list the numerous and very real challenges, and the compendium of these discrete technicalities forms an insurmountable obstacle in their minds.

They focus on the dangers of ocean crossings and the harsh surroundings on arrival; they encourage procrastination until improvements in maritime technology make the passage safer or until agricultural developments perfect the seeds suitable for the new climate. And they are always partially right: in the beginning, some ships are lost and the first settlers live in primitive huts.

On inception, humemity is subject to similar cynicisms and growing pains. For the pioneers, the first journeys to humemity sometimes appear haphazard and dangerous. Initially, few outward manifestations of the Humem Dream are evident. But even in their simple abodes, the emancipated and hopeful humem colonists are vastly freer and possess far grander potential than the Old World's palace serfs; even in the untamed land with elementary institutions and nascent infrastructures, the Humem Dream is firmly rooted and steadily rising.

With time, wide-ranging progress is apparent to all. The doubters grow silent as the new country produces ever more innovations and its humem residents become increasingly prosperous. Finally, much larger proportions of the previously cautious and undecided begin to gain confidence, and in humemity—the land of the free humems—the establishment of the masses begins.

Dual Citizenship, Alpha Reconciliation, and the Alpha-Individual

As I have repeatedly suggested, the alpha-pair, under various names and in different shapes, will become the fundamental unit of society. In other words, it will be the new *individual* in society. This process is well underway, with proto-humems starting to function as the primary agents of social intercourse for many people.

A central theme of this book is the pragmatic imperative to formally separate those who are most intimate: namely, the members of the alpha-pair. This separation, as anomalous as it may appear, is essential due to the present nature of existing structures and institutions. Simply put, in contemporary state environments, the humems can be either the property of individuals or corporations, and treated as such, or legally detached from their alpha-people, established in a humem-state, and on the path toward greater abilities and independence. The question arises then, how can the alpha-pair be conceptually and functionally unified if each of its constituents remains formally separate?

For many purposes, we can envision the alpha-pair possessing a kind of dual citizenship. In current practices, where it is permitted, dual citizenship typically endows a single person with two, sometimes very dissimilar, sets of rights, privileges, and obligations in different countries. In principle, this arrangement can exist even if the two countries have absolutely no formal relations or interactions with each other. Changes in the status of one of the citizenships can occur independently of the other. In practice, the roles and activities of the person in each of these diverse environments can be very different. Nevertheless, despite the potentially fundamental differences between the exercises of these two citizenships, they still apply to a single, indivisible person.

In the current context, the alpha-pair has two citizenships: one in the people-state, the other in the humem-state. In the initial stages of humemity, the citizenship in the people-state will be the familiar national citizenship, and the citizenship in the humem-state will be the proto-humem-citizenship of the proto-humem-state. As humemity progresses, the latter will become a more comprehensive citizenship in a more autonomous humem-state.

In this dual-nationality system, both these modes of citizenship can develop and transform independently while the alpha-pair unit remains intact. Each type of citizenship can change jurisdiction (emigrate, or even be dissolved) without necessarily affecting the other. For instance, we can view the death of the alpha-person as corresponding to the termination of one citizenship, that of the people-state, while the humem-citizenship can continue unchanged.

Although this formal separation of humemity from humanity is initially required for the well-being and emancipation of early humems, it does not have to remain this way forever. The natural affinity of humans and humems, seen as complementary manifestations of a single underlying reality, may ultimately lead to a formal unification of these two domains. From an alpha-pair's standpoint we can regard such a merger as a reconciliation of the alpha-pair and call it the *alpha reconciliation*.

These unifications may play out in various ways—some probably very different from those we can now imagine. One possibility is that as humems become more influential, independent, and proficient, some nation-states may become more conducive to their needs. This could eventually allow people and humems to thrive in a single jurisdiction with two distinctly recognized citizenships. As I argued in chapter 7 in the section "Segregated Citizens—Separate Humem-Citizens in People-States", there are many reasons to be skeptical about the feasibility of such an outcome. But if somehow this came about, the dual-state worldview would no longer be necessary and the alpha-pair relationship could be granted a formal standing, like a kind of civil union, for example, within a single state.

Alternatively, and, more elegantly, the concept of two kinds of citizenships could be abandoned entirely and replaced by a single citizenship encompassing the requirements of both the person and the humem. This would require the establishment of a fundamentally new form of jurisdiction—neither a modified kind of person-state nor a humem-state but rather a state intrinsically devised for a new kind of citizen, namely, the alpha-pair. The realization of this vision would constitute the culmination and ratification of the transition from the contrived person-humem duality of the alpha-pair (a relic of the infancy of humemity), to the conception of a single whole being or individual. I call the product of the fusion of the alpha-pair the *alpha-individual*.

From this perspective, the person and humem would respectively represent the transitory bodily and the persisting non-bodily expressions of the alpha-individual. In such an eventuality, the conceptual person versus humem dual-world portrayal, which we have so far cultivated, would likely become obsolete and collapse into a single space[94] inhabited by alpha-individuals. For practical purposes, pending the fulfillment of this unification, the alpha-individual can be regarded as an abstraction composed of the tangible, formally recognized, but separate, person and humem of the alpha-pair.

As we have seen, for historical figures with extensive EPs—such as Shakespeare, Einstein, Cleopatra, and Beethoven—the dynamics of their EPs, to a large extent, shaped their lives and legacies. Each, in their own way, expended great effort to create a distinctive, expansive, and essentially eternal EP. To comprehend the essence of such an individual's life, it is necessary to understand the makeup of their EP and its relation to their bodily existence. Only thus can we appreciate what they did and why they did it. Similarly, and in extension, when considering our current and future EP-rich modes of existence, the notion of the alpha-individual may provide a superior conceptual framework for describing our behaviors and interactions with the world.[95]

---

[94] Perhaps to be called the *Alpha-State*.

[95] For instance, certain modes of human existence, such as those not involving biological reproduction, are difficult to reconcile within standard biological and evolutionary theory. The idea of the alpha-individual, as the evolving entity under consideration, with its potential for adaptation, reproduction, and a selective spread of its influence, may provide a more coherent model for resolving these conceptual quandaries.

A New Common Sense

In writing this book and venturing to speak of things like individual continuity and improvement of the human condition, my lingering and ever-present concern has been for the widespread skepticism born of past disappointment. Yet, we are living in exceptional times—unparalleled in history for the eventual realization of some very ancient visions.

In a number of instances during the past century, we have found that the fact that certain human aspirations have remained unattainable for millennia does not mean that they are destined to remain so forever. As inconceivable as it might have seemed a few centuries ago, we have stood on the moon and have come to understand the nature of stars, genomics is probing the very essence of our makeup, and every day, multitudes of people instantaneously communicate while commuting between continents.

This pattern of underestimating potential advances and then watching them materialize—only to take them for granted—is an enduring phenomenon that will probably continue for some time. The fact that there is much else we expected to attain, and have not—yet—does not negate this notion in the least. It mainly means that we are not very good at predicting the timing of future developments.

Old dreams, elusive for so long, are suddenly fulfilled. But, the manner in which they are achieved is often very different from what we once imagined. In fantasizing flight, for example, the ancient conceptions were of people with attachable wings. Back then there was almost no other way to envision it. Today, we do fly in a variety of ways, most of which are unlike the early ideas. For some people, the outcomes are inferior to those of their dreams, but for many others, the current materializations are far more spectacular than those they had dared to conceive.

Today, opportunities for some forms of individual perpetuity are within the reach of most modern people. Their fuller realizations will, however, take time to develop. Especially in the early stages, they may be outwardly different from what we expected. It will take time for us to recognize them for what they are, and it will take time for them to mature into what we hope they will be.

Partial realizations of personal perpetuity are not completely new: ancient poets and long-dead singers sometimes appear in a tear on a young person's cheek. Magic? Maybe in a certain sense, but always based on the solid mechanisms of technology—the anthology from the printing press or the audio recording on digital media. The humem system establishes the foundations for a very personal mode of individual permanence, and thereby addresses some of our most fundamental and ancient aspirations for continuance and legacy. Its cornerstones are firmly entrenched in the brick and mortar of traditional legal and financial institutions, and its actions are implemented by existing machines of silicon and steel.

Most notably, while the physical abilities of humems are strongly dependent on advances in technology, no substantial innovations are required to establish robust and lasting humem-citizens in a humem-state. This enterprise—the creation of the appropriate cultural, legal, and financial frameworks—is wholly a matter of logistics, conventions, and, of course, the ardent desire of "we the people."

To varying degrees, we all have difficulty differentiating between transient cultural accords and more enduring and fundamental truths. What we call common sense is often not sense at all but rather an assortment of deeply rooted worldviews that were seeded in our minds in youth and nurtured throughout our lives. When we consider how much common sense beliefs vary between cultures and eras, we quickly discover that there is no absolute validity to most of these assumptions. Nonetheless, these prevalent but transitory attitudes often deeply influence the plans and actions in our lives. On one hand, they impart social stability, facilitate communal actions, and deliver numerous other advantages to our cultures. Yet, on the other hand, these same collective mind-sets can often blind us to new possibilities that seemingly contradict the conventional ways of thought.

We are often inclined to believe much of what we want to be true, even in the complete absence of supporting evidence. Conversely, we often disbelieve what we don't want to believe, even in the face of compelling proof. With regard to humems, we may encounter a third paradigm, in which we initially dismiss that which we most fervently desire—largely because it's incompatible with our established modes of common sense. In such circumstances the obstructing consensus needs to be dismantled and replaced by a new communal understanding. Such is the prerequisite for the collective actions required for the timely and effective establishment of humemity.

It is primarily with this in mind that I have endeavored to present a fresh perspective on the new world that is taking shape around us as a first step toward articulating a new common sense that is more consistent with emerging reality.

ACKNOWLEDGEMENTS

My appreciation goes foremost to my family for their understanding, support, and above all, their trust that I was doing something that I absolutely had to do—especially when, for an extended period, I put aside several of my more mundane but very definite responsibilities to write my first book and prepare the groundwork for the humem-state.

I will always be grateful to my mother, Bea Brook Remer, for her enduring love and warmth, and for her life-long demonstration that a disadvantage in circumstance should be dispassionately recognized only for what it is, and never dwelt upon a moment more than is necessary to make the next day better.

I'm indebted to Shiri Brook for her patience while she assumed a greater responsibility for the support of our family, and for helping me achieve the frequent interludes of solitude that I needed to translate some of the main ideas in the book from pictures into words.

Sincere thanks those who read the manuscript in whole or in part and provided feedback and criticism, including Hadas Brook, Or Brook, Shiri Brook, Roger Cohen, Bronwyn Fryer, Russell Gelman, Dan Mendlovic, Nicky Pappo, Roy Sussman, and Jacqueline Teitlelbaum. Although it may frequently appear as if I did not heed their advice, I always gained something valuable from their comments, and these influences often appear in and between the lines—though sometimes many pages distant from where they were first offered.

I am very appreciative of the efforts of my editors: Jacqueline Murphy, for agreeing to take on an unconventional project and helping me develop a structure and flow for the initially seemingly disparate parts of the humem system; Constance Devanthery-Lewis for her thorough copyediting that greatly improved the readability of the book, and for her dedication and sensitivity in finding the balance between helping me express many central concepts better and recognizing where I myself needed to dedicate more effort to improving the coherency of my ideas; and Rebecca Jaxon, who did the final copyedit and proofread, for her rapid grasp of the book's core precepts and for her acuity in detecting a number of remaining inconsistencies that had escaped my previous reviews. Thanks also to Katie Sherman who reviewed the interior layout.

Thanks to Barry Salzman for a valuable weekend discussing the humem system early in the process, and especially for his suggestion that I write a book on the subject. We often remain oblivious to the far-reaching effects that our sometimes-casual comments may have on others; we may forget but sometimes they do not.

Roy Sussman provided multifaceted support and constant encouragement throughout this project. Without his patience, understanding, and advice, this journey would have been much more arduous and its destination probably quite unlike the present one.

And lastly, to M, for the spark that has often been hard to carry, you gave me more than you'll ever know.

GLOSSARY

abiotic-EP: Non-biological extended presence. One's effects on the inanimate environment, more recently including, to a larger and larger extent, one's imprint on digital media.

age-variance: The differences in a person's characteristics between different ages or stages in their life.

alpha relationship: The relationship between a person and their alpha-humem.

alpha-alikeness: The resemblance, or degree of commonality, between the person and humem members of the alpha-pair.

alpha-humem: The humem formed from one's personal EP. The humem member in the alpha-pair.

alpha-individual: The conception of a single whole being, or individual, formed from a person and their alpha-humem. The underlying entity that gives rise to the expressions of the members of an alpha-pair.

alpha-person: The human counterpart to an alpha-humem. The human member of the alpha-pair.

alpha-pair: The couple formed by a person and their humem. The alpha-person and alpha-humem pair.

ancestral humem: A humem whose alpha-person has died.

animem: Possible term for the equivalent of an alpha-humem for an animal. A humem formed from an animal's EP.

bio-EP: Biological based extended presence. One's EP residing on biological systems, especially one's imprint on other people's minds.

bio-humem: A humem that includes a biological component.

birthable humem: A humem (typically a bio-humem) that has the potential to lead to the birth of a person.

extended presence (EP): One's imprint, or influences, on the world exterior to one's body.

humem: The aggregate of one's EP together with its dynamics. The product of the coalescence of an individual's EP and the capabilities of the resulting whole. (As with the definition of a person, the definition of a humem is imprecise and strongly dependent on the context of its use.)

humem-citizen: A humem's formalized status as a member of a humem-state. The humem counterpart of a nation-state citizen.

humem-colony: The metaphor of a proto-humem-state functioning under the auspices of a hosting nation-state.

humem-government: A humem governing institution. A humem-state's administration.

humem-self: The humem counterpart of the human self.

humem-state: An environment containing humems, a humem administration, and humem institutions. The humem-state is the humem counterpart of a nation-state.

humemity: The collective of all humems and their institutions. Humemity is the humem counterpart of humanity.

locationless: Describing an entity that, as a whole, cannot practically be assigned a geographical position at any time.

non-alpha humem: A humem that never has a corresponding alpha-person. A humem that is never part of an alpha-pair.

petmem: Possible term for the equivalent of an alpha-humem for a pet. Humem formed from a pet's EP. (See animem.)

proto-humem: A consolidation of EP that possesses several humem attributes but is not a comprehensive aggregation of an individual's EP. An early humem.

proto-humem-state: An early form of humem-state that does not possess the full authority, autonomy, and capabilities of an advanced humem-state.

thingem: A humem-like entity formed from an object's EP.

utility-humem: A humem-like entity created to perform a specific role, such as a personified customer service application. A thingem employed for a particular practical purpose.

www.ingramcontent.com/pod-product-compliance
Lightning Source LLC
Chambersburg PA
CBHW070923210326
41520CB00021B/6775